SpringerBriefs in Pharmaceutical
Science & Drug Development

For further volumes:
http://www.springer.com/series/10224

Gabriela Hrckova · Samuel Velebny

Pharmacological Potential of Selected Natural Compounds in the Control of Parasitic Diseases

 Springer

Gabriela Hrckova
Slovak Academy of Sciences
Institute of Parasitology
Kosice
Slovakia

Samuel Velebny
Slovak Academy of Sciences
Institute of Parasitology
Kosice
Slovakia

ISSN 1864-8118
ISBN 978-3-7091-1324-0
DOI 10.1007/978-3-7091-1325-7
Springer Wien Heidelberg New York Dordrecht London

ISSN 1864-8126 (electronic)
ISBN 978-3-7091-1325-7 (eBook)

Library of Congress Control Number: 2012952683

Printed on acid-free paper

Springer is part of Springer Science+Business Media (www.springer.com)

Preface

The disease burden caused by parasitic protozoa and worms is significant and represents a challenging health problem mainly in tropics. The drugs currently used for the treatment of parasitic infections are mostly effective, however, some of them have limitations, such as toxic side effects and high cost. Moreover, some parasitic organisms developed resistance to many of these drugs. There is a need for discovery of new drugs. For centuries, medicinal plants have been used to combat parasitism, and in many parts of the world are still used for this purpose. Plants are valuable sources for the screening of bioactive secondary metabolites, but also bacteria, fungi, terrestrial and marine invertebrates produce pharmaceutically useful compounds with potential antiparasitic activity. Natural product research shows promise in finding new lead structures besides rational drug design.

The search for bioactive natural molecules begins with the screening of various extracts, isolation of active fractions, and identification of the active components when possible. This search must consider, among other things, target specificity, making better use of the biochemical and biological characteristics of individual parasite species, and cytotoxicity determination. This is a multidisciplinary process, which is initially realized in vitro, however, therapeutic concepts have to be validated and toxicity of selected compounds evaluated in animal studies.

Malaria, trypanosomiasis, and leishmaniasis are among the most important public health problems in developing countries. Chapter 1 reviews the most promising results obtained during the screening of antimalarial, trypanocidal, and leishmanicidal properties of natural products classified according to their chemical structure since the year 2000.

Chapter 2 is devoted to the review of anthelmintic properties of bioactive compounds isolated from the various natural sources. Particular attention is paid to the mode of anthelmintic action of natural alkaloids, essential oils, flavonoids, glycosides, saponins, condensed tannins, endoperoxide sesquiterpene lactones, enzymes, and amides against many species of round and flatworms.

How do the natural compounds exert their anthelmintic effects? To get the answers to this question, many examples are presented from the point of view of benefits for the host (antioxidant and immunomodulatory effects) and direct anthelmintic action of screened compounds in Chap. 3.

It is our hope that this book will provide the readers with a useful survey of the recent results in the screening of natural compounds with promising antimalarial, trypanocidal, leishmanicidal, but mainly anthelmintic effects and that it may serve as a valuable source of information for scientists, postgraduate and graduate students working in medical and pharmacological research.

Acknowledgments

We would like to express our deep gratitude to Prof. Chris Arme (Keele University, United Kingdom) for his essential help in the manuscript preparation. We also thank for financial support to the Scientific Grant Agency of the Ministry of Education of the Slovak Republic - VEGA, under contract Nos. 2/7188/27 and 2/0188/10.

Contents

Introduction

For thousands of years, natural products have played an important role throughout the world in treatment and prevention of various diseases in humans and animals. Natural product medicines have been collected from various sources, mostly terrestrial plants and microorganisms and marine organisms, and they have been the major source of chemical diversity for starting materials driving pharmaceutical discovery over the past century.

Infectious diseases caused by parasitic protozoan and metazoan species, in spite of an immense effort put into eradication programs within the past 20 years, are still on the list of the most frequently encountered diseases in tropical countries. The most debilitating protozoan diseases, including malaria, leishmaniasis, and trypanosomiasis, for millenia have prevented economic and cultural development in vast regions of the world and still are a major social, economical, and health problem today. Estimates show that approximately 300–500 million people are at risk from these infections. The World Health Organization estimates that approximately 2 billion people harbor parasitic worm infections, and worm infections in livestock are an important factor affecting food production.

Before commercial anthelmintics were introduced into the world market, worm infections were controlled using specific plants, which, based more on belief rather than on sound knowledge, were credited with having specific actions. Plants with antiparasitic properties could be found in temperate, tropical as well as colder climates in the world; however, the highest diversity of plants with ethnomedicinal or ethnoveterinary records can be found in tropical regions. Nowadays, therapy and prevention of parasitic infections is facing several major issues: emergence and rapid spread of resistant strains of parasites and the limited number of safe and highly effective antiparasitic drugs, which in turn leads to monotherapy of a particular disease.

The frequent use of a few alternative drugs, together with other biological factors, contributed to the development of multidrug resistant malaria strains with no alternative treatment possibilities. There are only a few classes of anthelmintic drugs for therapy and prevention of so-called neglected tropical diseases of humans caused by metazoan parasites, and drug resistance has been reported already from some endemic regions. Anthelmintic resistance of gastrointestinal nematodes in small ruminants, which are bred for the production of meat, milk, or wool, is a global problem that leads to enormous economic losses. In this context, the administration of combinations of either two anthelmintics or co-administration of drugs, with natural compounds with a similar spectrum of activity and different mechanisms of action, has been suggested as a potential means of delaying the development of drug resistance.

The very high cost of discovering and developing new drugs has led to the introduction of only a few new products in the market displaying a new mode of action within the past 20 years. To identify new antiparasitic lead compounds, large numbers of compounds will have to be examined in pre-clinical tests. They come mostly from the laboratories of medical chemists, but highly successful sources of compounds can also be found in nature. Higher plants, but also symbionts like lichens, are able to synthesize the specific compounds identified as secondary metabolites. As the term implies, they are not strictly essential to the main functions of the plants, such as their growth and reproduction. They have been associated, for example, with plant defence mechanisms against pathogens, but other functions have also been described. Plant-derived secondary metabolites can be divided, on the basis of their molecular formulae and structural motifs, into several classes and the most abundant are essential oils, flavonoids, alkaloids, saponins, glycosides, tannins, sesquiterpene lactones, lactones with peroxidic structure, amides, and proteins with enzymatic activity.

The WHO has recognized the importance of traditional medicine since the late 1970s, when its Traditional Medicine Programme was established. It has developed guidelines for the assessment of herbal medicines focussing mainly on malaria and other important protozoan infections. Similar programs have been initiated since the 1980s to discover novel lead molecules against filarial nematodes, human trematodiasis, and gastrointestinal nematodes of livestock. Isolation of artemisinins from the plant *Artemisia annua* and avermectins from the terrestrial actinomycete, *Streptomyces avermectinius*, within the 1970s are probably two major successes of intensive screening program of compounds from natural sources. The world's ocean is extremely rich in living organisms, of which marine invertebrates comprise a diverse and promising source of compounds from a wide variety of structural classes. Marine-derived small molecules have been described from marine plants, animals, algae, fungi, and bacteria, etc. Many of them are chemically unique compounds and many have been successfully isolated and characterized, and synthetic derivatives have been subjected to a large-scale screening against protozoan parasites.

Since 2000, the interest in evaluation of biological, medicinal, and antiparasitic activities of natural compounds has increased considerably, as indicated by the high number of published scientific papers. After reviewing numerous papers, we noticed that a very high number of small compounds from natural sources were evaluated for their cytotoxic activity against three major human protozoan infections, with a large proportion of molecules being derived from marine organisms. On the other hand, antiparasitic effects against trematodes, cestodes, and nematodes were attributed mostly to either the whole plant extracts or to a whole group of secondary metabolites, like tannins or essential oils. Lower numbers, and mostly very recent studies, involved chemically defined, and either purified or chemically synthesized, natural compounds using a variety of experimental helminth infections. Therefore the main purpose of this book was to draw attention to the wide variety of compounds from various natural sources, so far evaluated for antiparasitic activity. We also aimed to bring an overview of their potential not only against major protozoan infections of humans, but also against the most pathogenic helminth parasites together with the possible mechanism of action, cytotoxicity, or side effects on hosts. We hope that this comprehensive overview will illustrate the complexity of this issue for those interested in pharmacological research in the field of parasitology.

Chapter 1
Pharmacological Potential of Natural Compounds in the Control of Selected Protozoan Diseases

Abstract Malaria, trypanosomiasis, and leishmaniasis (neglected tropical diseases), belong to the most devastating diseases affecting humans and animals in developing regions of Asia, Africa, and South America. The drugs, currently used for the treatment of these diseases, are mostly effective; however, some of them have limitations, such as toxic side effects and high cost. Moreover, *Plasmodium*, *Trypanosoma*, and *Leishmania* have developed resistance to many of these drugs. In the first part of this chapter, problems of pathology, therapy, and drug resistance are reviewed. The following part focuses on plants, bacteria, fungi, and marine organisms, which provide invaluable sources of compounds with antiparasitic potential. The compounds isolated from nature are classified according to their chemical structure, and methods for evaluation of their antiparasitic activity are also discussed. The review of promising results, obtained by many investigators from the year 2000 by screening of natural compounds evaluating their antiplasmodial, trypanocidal, and leishmanicidal activity, are presented in the last three parts of this chapter.

Keywords Malaria · Trypanosomiasis · Leishmaniasis · Pathology · Current drugs · Drug resistance · Drug discovery · Higher plants · Marine organisms · Secondary metabolites · Screening of natural compounds

1.1 The Most Serious Protozoan Infections: Pathology, Therapy, and Drug Resistance

Protozoan diseases are the oldest and most devastating tropical diseases affecting humans and animals. They are responsible for considerable mortality throughout the world but are predominant in the tropics and subtropics (WHO 2002). Malaria,

G. Hrckova and S. Velebny, *Pharmacological Potential of Selected Natural Compounds in the Control of Parasitic Diseases*, SpringerBriefs in Pharmaceutical Science & Drug Development, DOI: 10.1007/978-3-7091-1325-7_1, © The Author(s) 2013

trypanosomiasis, and leishmaniasis are protozoan diseases caused by parasites of the genus *Plasmodium, Trypanosoma,* and *Leishmania,* respectively. They exhibit a combination of biochemical, physiological, nutritional adaptation and have a tendency to induce immunity in their hosts.

Malaria is the most serious and widespread parasitic disease encountered by mankind. Every year, about 500 million people are afflicted and about 2.7 million people 5 years old and pregnant women due to lack of or low immunity (WHO 2002). Human infection can be caused by four distinct species of the genus *Plasmodium,* but *P. vivax* and *P. falciparum* account for more than 95 % of malaria cases. Malaria is transmitted by the bite of the female anopheles mosquito, at which time sporozoites of *P. falciparum* are discharged into the puncture wound. Sporozoites are then carried to the liver, where they enter hepatic mesenchymal cells and begin to grow. Lysis of the hepatocyte then releases the merozoite form of *P. falciparum,* which invades host red blood cells (RBCs), feeding on hemoglobin during the erythrocytic portion of its life cycle. In the RBC, parasite expresses a number of polypeptide products that are exported to the surface of the RBC, rendering it antigenic. In order to escape the host immune system, the parasite regularly exchanges these peptides in a process called antigenic variation. Malaria has become difficult to treat, due to an increase in multidrug resistant strains. Current major areas of antimalarial research include prevention and vector control, identification of new targets, development of an effective vaccine, investigation of selected medicinal plants, and design and synthesis of new antimalarial agents (Woster 2009; Muthaura et al. 2011).

The cinchona alkaloid *quinine* (isolated from *Cinchona succiruba* bark, Rubiceae) was the first compound which exhibited significant antimalarial activity. Subsequent studies produced the synthetic analog chloroquine, which was initially an excellent treatment for malaria. The evolution of chloroquine-resistant strains of *P. falciparum* has rendered this drug useless in certain areas of the world. Amodiaquine, tebuquine, tafenoquine, primaquine, and mefloquine are synthetic analogs related to quinine and chloroquine. The mechanism of all 4-quinoline antimalarials appears to be the same. During the degradation of hemoglobin, *Plasmodium* must eliminate free heme, which is a toxic by-product. The parasite forms a polymer of heme known as hemazoin to alleviate the toxicity of heme. It is thought that the 4-quinolines cap the growing hemazoin polymer, and the parasite is killed by heme toxicity (Woster 2009).

Perhaps the most promising advance in the treatment of malaria is the discovery of *artemisinin* (Chinese name: qinghaosu) from *Artemisia annua* (Asteraceae), an annual herb that has been used in traditional Chinese medicine for over 2,000 years (van Agtmael et al. 1999). Artemisinin is hydrophobic, passes biological membranes easily, and is a potent antimalarial with an IC_{50} value of 7.3 nM against *P. falciparum.* This 1,2,4-trioxosequiterpene lactone produces oxidative stress in *P. falciparum* by an endoperoxide group that is essential for its activity (Woster 2009). Artemisinin generates bioreactive radicals capable of intracellular damage, and inhibits the *Plasmodium falciparum* endoplasmic reticulum calcium pump (SERCA) (Eckstein-Ludwig et al. 2003). Artemisinin resistant *P. falciparum* contains SERCA mutations (van Agtmael et al. 1999; Woodrow et al. 2005). Artemisinin derivatives have been used to treat malaria around the world, and their extensive usage has not been associated with any significant toxicity (Taylor and White 2004). IC_{50} values range from 4.2 to 16.2 nM for different derivatives (Woodrow et al. 2005). WHO (2001) recommends combinations of drugs containing artemisinin to overcome the resistance to conventional monotherapies.

Antimalarial drug resistance has been defined as "the ability of a parasite to survive and/or multiply despite the administration and absorption of a drug given in doses equal to or higher than those usually recommended but within tolerance of the subject. The drug in question must gain access to the parasite or the infected red blood cell for the duration of the time necessary for its normal action" (EAC 2012).

A multitude of factors can be involved in treatment failure including problems with non-compliance and adherence, poor drug quality, interactions with other pharmaceuticals, poor absorption, misdiagnosis, and incorrect doses being given. The majority of these factors also contribute to the development of drug resistance. It is generally accepted to be initiated primarily through a spontaneous mutation. The genetic events that confer antimalarial drug resistance are spontaneous. The generation of resistance can be complicated and varies between *Plasmodium* species.

Resistance has emerged to all classes of antimalarial drugs except the artemisinins. The biological mechanism behind the resistance to chloroquine is related to the development of an efflux mechanism that expels this drug from the parasite before the level required to effectively inhibit the process of heme polymerization, which is necessary to prevent buildup of the toxic products formed by hemoglobin digestion. The resistance of other quinolone antimalarials such as amodiaquine, mefloquine, halofantrine, and quinine are thought to have occurred by similar mechanisms.

Plasmodium has developed resistance against antifolate combination drugs, the most commonly used being sulfadoxine and pyrimethamine. Two gene mutations are thought to be responsible, allowing synergistic blockages of two enzymes involved in folate synthesis.

A number of behavioral, pharmaceutic, and pharmacokinetic factors affect the probability of parasites encountering subtherapeutic levels of antimalarial agents. Several antimalarial drugs (notably lumefantrine, halofantrine, atovaquone, and, to a lesser extent, mefloquine) are lipophilic, hydrophobic, and quite variably absorbed. Poor oral bioavailability, with a consequent wide range in blood levels, will favor the emergence of resistance. Improving oral bioavailability thus reduces doses required to clear infection and should reduce the emergence and spread of resistance (White 2004).

Trypanosomiasis is another highly disabling and fatal disease, common in Africa and Latin America. The African trypanosomiasis (sleeping sickness) is found only in Africa where about 50,000 new cases are reported annually. Its treatment is still a challenge due to marginal efficacy and severe toxicity of available drugs, therefore, a need for search of new less toxic and more effective drugs. Trypanosomiasis is a group of diseases in vertebrates caused by trypanosomes of the genus

Trypanosoma. Trypanosomes are a group of kinetoplastid protozoa, distinguished by having only a single flagellum (Mwangi et al. 2010).

African trypanosomiasis, also known as *sleeping sickness,* is caused by two morphologically identical parasites. *T. brucei rhodensiense* produces an acute form of the disease known as East African trypanosomiasis, while *T. brucei gambiense* produces a chronic disease called West African trypanosomiasis. Sleeping sickness is transmitted by tsetse flies of the genus *Glossina.* The metacyclic trypomastigotes rapidly transform into bloodstream trypomastigotes within the extracellular spaces in the subcutaneous tissue of the host. Wild animals and cattle are important reservoir hosts for *T. b. rhodensiense,* while humans are the main reservoir for *T. b. gambiense.* In the first stage of the disease, trypanosomes enter the bloodstream and multiply there. The second stage involves central nervous system (CNS) invasion. The parasite is protected against oxidative stress by reduced trypanothione, a metabolite of polyamine metabolism in trypanosomes. In the presence of oxidative stress, oxidized trypanothione is formed, and must be recycled to the reduced form by a trypanothione reductase, an enzyme unique to the parasite (Woster 2009).

Early stage disease (when parasites are largely found in the blood) is treated effectively with suramin and pentamidine Suramin, a polysulfonated urea, is thought to slowly enter the parasite by receptor-mediated endocytosis, which is possibly linked to host LDL-endocytosis. Pentamidine is only used when therapy with suramin is contraindicated. There are only two effective treatments for late stage trypanosomiasis—melarsoprol and eflornithine. These agents must penetrate the CNS in order to be effective. The mechanism of action of melarsoprol could be the combination of trypanothione depletion and inhibition of trypanothione reductase. Eflornithine is considered for the treatment of arsenic-(and hence melarsoprol) resistant *T. b. gambiense* causing sleeping sickness (Woster 2009; Varughese et al. 2010).

American trypanosomiasis (Chagas disease) is caused by *T. cruzi* and is a major health problem in South America. This zoonotic infection is transmitted by bloodsucking triatomine insects, when the bite wound is contaminated with insect feces containing the non-replicative trypomastigote form of the parasite. Once inside the host, trypomastigotes enter any nucleated cells and transform into the replicative amastigote form that transforms into trypomastigotes. These are freed to infect nearby cells and to enter the circulation, which disseminates the infection in the host. The parasite life cycle is completed when the triatomine insect sucks host infected blood. The ingested trypomastigotes transform into the replicative epimastigote form in the insect midgut (Burleigh and Woolsey 2002).

Chagas disease is characterized by three clinical phases—acute, intermediate, and chronic, that differ in symptoms and morbidity. *T. cruzi* is much more difficult to treat than *T. brucei gambiense* or *T. brucei rhodensiense,* since this trypanosomatid parasite is intracellular and drugs used for the disease must pass through mammalian and parasite cell membranes to be effective. Two drugs are currently approved for treatment of Chagas disease—nifurtimox (derived from nitrofuran) and benznidazole (a nitroimidazole derivative). The use of these drugs to treat the acute phase of the disease is widely accepted. However, their use in the treatment of the chronic phase is controversial (Salas et al. 2011).

Some trypanocidal drugs such as nifurtimox and benznidazole act through free radical generation during their metabolism. *T. cruzi* is very susceptible to the cell damage induced by these metabolites because enzymes scavenging free radicals are absent or have very low activities in the parasite. Another potential target is the biosynthetic pathway of glutathione and trypanothione, the low molecular weight thiol found exclusively in trypanosomatids. These thiols scavenge free radicals and participate in the conjugation and detoxication of numerous drugs. Inhibition of this key pathway could render the parasite much more susceptible to the toxic action of drugs such as nifurtimox and benznidazole without affecting the host significantly. Other drugs such as allopurinol and purine analogs inhibit purine transport in *T. cruzi*, which cannot synthesize purines de novo. Nitroimidazole derivatives such as itraconazole inhibit sterol metabolism. The parasite's respiratory chain is another potential therapeutic target because of its many differences with the host enzyme complexes. A large set of chemicals of plant origin and a few animal metabolites active against *T. cruzi* are enumerated and their likely modes of action are briefly discussed in the review of Maya et al. (2007).

Drug resistance in trypanosomiasis. Current chemotherapy treatment is complex, since special drugs have to be used for the different development stages of the disease, as well as for the parasite concerned. Melarsoprol is the only approved drug for effectively treating both subspecies of human African trypanosomiasis in its advanced stage, however, the drug's potency is constrained due to unacceptable side effects (encephalopathy). In addition to the deleterious treatment with melarsoprol, the number of drug resistant strains of *T. brucei supp.* increases. Mechanisms of drug resistance have been elucidated and involve decreased drug import through the loss of the purine transporter P2 as well as enhanced drug export, mediated by a multidrug resistance-associated protein named TbMRPA (Wilkinson and Kelly 2009). Thus, the medical treatment with the available chemotherapeutics becomes exceedingly difficult (Gehrig and Efferth 2008). Eflornithine has replaced melarsoprol in many regions. However, a need for protracted and complicated drug dosing regimes slowed widespread implementation of eflornithine monotherapy (Barret et al. 2011). A promising strategy for research into new drugs and moreover, to overcome drug resistance, are compounds derived from natural sources.

Leishmaniasis afflicts large areas of Africa, Asia, Mediterranean, and Latin America (WHO 2002). Leishmaniasis is a disease complex caused by 17 different species of protozoan parasites belonging to the genus *Leishmania* (Croft et al. 2006). The parasites are transmitted between mammalian hosts by sand flies of the genera *Lutzomia* and *Phlebotomus* (Rocha et al. 2005). Parasites, inoculated by the vector as the flagellate promastigotes, enter the mammalian host, where they infect macrophages, differentiating into non-motile amastigotes and multiplying as such. The disease can be divided into three clinical manifestations—cutaneous, mucocutaneous, and visceral leishmaniasis. Cutaneous and mucocutaneous leishmaniasis are diseases that are generally not life threatening. Visceral leishamaniasis (also known as *kala-azar* in India or *dumdum fever* in Africa) produces life threatening systemic infection if left untreated.

In general, the intracellular amastigote form is considered the target for chemotherapy of leishmaniasis. The standard *treatment* for visceral leishmaniasis is sodium stibogluconate or meglumine antimoniate, moderately toxic pentavalent antimonials for which resistance is increasing. The mechanism of the antileishmanial activity of antimony remains to be elucidated, but it is known that these drugs suppress glycolysis and fatty acid metabolism in the parasitic glycosome. Other drugs for leishmaniasis are pentamidine, amphotericin B, miltefosine (Varughese et al. 2010; Woster 2009). These drugs currently used for treatment of leishmaniasis have limitations due to toxicity, route of administration, expense, and due to resistance developed by *Leishmania* spp. Hence, there is a need to develop new effective drugs with reduced toxicity and affordable prices. Some metabolic pathways, which are essential and could be used as potential drug targets, are discussed in the review of Chawla and Madhubala (2010). Trypanothione pathway and topoisomerases are frequently discussed by investigators searching for new leishmanicidal drugs. The review, presented by Das et al. (2004), summarizes topoisomerase genes and proteins of kinetoplastid parasites *Leishmania* and *Trypanosoma* and the roles of these enzymes as targets for therapeutic agents.

Drug resistance in leishmaniasis. For almost seven decades pentavalent antimonials constituted the standard antileishmanial treatment worldwide. However, over the past 15 years their clinical value decreased due to the widespread emergence of resistance to these agents (Maltezou 2010). Reduction of drug concentration within the parasite, either by decreasing drug uptake or by increasing efflux/sequestration of the drug, constitutes the primary mechanism of antimonial resistance; other potential resistance mechanisms include inhibition of drug activation, inactivation of active drug, and gene amplification (Kothari et al. 2007; Mittal et al. 2007). Variation in the efficacy of drugs in the treatment of leishmaniasis is frequently due to differences in drug sensitivity of *Leishmania* species, the immune status of the patient, or the pharmacokinetic properties of the drug (Croft et al. 2006).

Thiol metabolism possesses a key role in both clinical- and laboratory-generated resistance mechanisms. In natural antimonial resistance, the impaired thiol metabolism results in inhibition of SbV activation and decreased uptake of the active form SbIII by amastigotes. It has been suggested that decreasing the intracellular thiol concentrations through thiol depletors may increase the leishmanicidal action of drugs and thus reverse parasite resistance (Mittal et al. 2007).

Overexpression of the membrane-bound ATP-binding cassette (ABC) transporters on the surfaces of leishmania parasites is another mechanism of antimonial resistance. In addition to leishmanias, this transport system modulates the efflux and intracellular accumulation of various drugs and thus resistance in other parasites (e.g., *Plasmodium* spp.) and also in cancer cells. Overexpression of ABC transporters concerns laboratory-derived and in-field resistant parasites (Mukherjee et al. 2007). Flavonoid dimers are also known to reverse antimonial resistance in leishmanias in vitro by inhibiting ABC transporters and increasing the intracellular accumulation of the drug (Wong et al. 2007). Understanding their molecular and biochemical characteristics will lead to the design of new drugs because the development of antileishmanial is generally slow (Maltezou 2010).

1.2 Bioactive Natural Compounds: Their Chemical Classification and Evaluation of Biological Activity Against *Plasmodium, Trypanosoma,* and *Leishmania*

Traditional societies have used a number of medicinal plants to treat parasitic infections since ancient times. *Ethnomedicinal plants*, (i.e., plants indigenous to populations, specific cultural or ethnic groups) being locally available, the therapeutic preparations from these plants are cheap, cost–effective, and easily affordable. Traditional wisdom is a base on which one can rely upon in the selection of plants for drug prospecting. This strategy helped researchers in the discovery of a number of new drugs and candidates for synthetic modification in medicinal chemistry with an aim to create safer compounds with reduced toxicity, reduced side effects, or improved bioavailability. The challenge will be to translate herbal medicine practice into an evidence-based monotherapy or combined therapy. Plants traditionally used by different cultures may provide cheap and effective alternatives to the currently used drugs. These plants have offered many substances that are claimed as "lead molecules" for synthesis of new classes of antiprotozoal chemotherapeutic drugs.

Secondary metabolites in plants are organic compounds that are not directly involved in the normal growth, development, or reproduction of an organism. Unlike primary metabolites, absence of secondary metabolites does not result in immediate death, but rather in long-term impairment of the organism's surviving and reproductive potential, or perhaps in no significant change at all. Secondary metabolites are often restricted to a narrow set of species within a phylogenetic group. The chemical composition of the various plant constituents is affected by the climatic conditions and the locality under which the plant species are growing (Muthaura et al. 2011). Secondary metabolites often play an important role in plant defence against herbivory and other interspecies defences. Humans use secondary metabolites as medicines, flavorings, and drugs.

Secondary metabolites can be classified on the basis of chemical structure (for example, having rings, containing a sugar), composition (containing nitrogen or not), their solubility in various solvents, or the pathway by which they are synthesized (e.g., phenylpropanoid, which produces tannins). A simple classification includes three main groups: the terpenes (made from mevalonic acid, composed almost entirely of carbon and hydrogen), phenolics (made from simple sugars, containing benzene rings, hydrogen, and oxygen), and nitrogen-containing compounds (extremely diverse, may also contain sulfur). Excellent review of plants' secondary metabolites as potential drugs or as leads against protozoan diseases has been presented by Schmidt et al. (2012a, b).

The traditional cultures sometimes used essential oils in treatment or prophylaxis (Kaur et al. 2009). An essential oil is a concentrated hydrophobic liquid containing volatile aromatic compounds from plants. Essential oils are also known as volatile oils, aethereal oils or aetherolea, or simply as the "oil of" the plant from which they were extracted, such as *oil of clove*. An oil is "essential" in the sense that it carries a distinctive scent, or essence, of the plant. Essential oils do not form a distinctive category for any medical, pharmacological, or culinary purpose. Two separate modes of action can be attributed to the efficacy of plant essential oils for treating parasitic infections: their immunomodulatory properties and their antiparasitic effects (Anthony et al. 2005).

Kayser et al. (2003) classified natural products as potential antiparasitic drugs from different sources (plants, fungi, bacteria, and marine products) into three main groups:

Alkaloids—quinolines, bisbenzylquinolines, benzyl- and naphthylisoquinoline, indole.
Terpenes—sesquiterpenes, diterpenes, limonoids.
Phenolics—lignans, chalcones and aurones, flavonoids, naphthoquinones, and related quinones.

Natural products as potential antiprotozoal drugs can be classified in more detail (Kayser et al. 2002): *Lipids and related aliphatics*—1. organic acids, lipids, and acetogenins, 2. polyenes, 3. ajoenes. *Phenols*—1. simple phenols, phenolic acids, and coumarins, 2. lignans, 3. chalcones, and aurones, 4. flavonoids, 5. naphthoquinones, 6. anthraquinones and xanthones. *Terpenoids*—1. iridoids, 2. monoterpenes, 3. sesquiterpene lactones, 4. diterpenes, 5. triterpenes, 6. limonoids, 7. quassinoids. *N-Containing natural products (non-alkaloids)*—1. steroidal alkaloids, 2. other N-containing compounds. *Alkaloids*—1. quinolines, 2. benzyl- and naphtylisoquinoline alkaloids, 3. bisbenzylisoquinolines, 4. indoles. *Other natural product classes*—1. nucleosides, 2. aminoglycosides.

Marine sponges and associated microbes have been reported to produce therapeutically important compounds. These sponges are simple multicellular invertebrates attached to solid substrates in benthic habitats. Most of the species are placed under the class Demospongiae. Since sponges are simple and sessile organisms, during evolution they have developed potent chemical defensive mechanism to protect themselves from competitors and predators as well as infectious microorganisms. Studies show that secondary metabolites in sponges play a crucial role in their survival in the marine ecosystem. These natural products have interesting biomedical potential, which is attributed to their antiviral, antitumor, antimicrobial, and general cytotoxic properties. The chemical diversity of secondary metabolites isolated from sponges includes amino acids, nucleosides, macrolides, porphyrins, terpenoids, aliphatic cyclic peroxides, and sterols. Sponges are well known to be hosts for a large community of microorganisms. It has been recognized that the sponge-associated bacterial community consists of at least ten bacterial phyla. Other symbiotic microbial populations that inhabit sponges are fungi and microalgae. It is hypothesized that symbiotic marine microorganism harbored by sponges are the original producers of these bioactive compounds (Zhang et al. 2005). As infectious microorganisms evolve and develop resistance to existing pharmaceuticals, marine sponges provide novel leads against bacterial, viral, fungal, and parasitic diseases (Laport et al. 2009). In the review of Thomas et al. (2010) an effort has been made to relate the biomedical significance of secondary metabolites of sponge-microbial association, which were discovered so far.

Screening methods for finding antiprotozoal potential of natural compounds are usually standardized and well described. Dua et al. (2011), Mwangi et al. (2010), Malebo et al. (2009), Sanchez et al. (2006), Carvalho and Ferreira (2001), and many other investigators described the in vitro methods for evaluation of antiplasmodial, antitrypanosomal, and antileishmanial activity, as well as cytotoxicity.

Many overviews are restricted only for determination of biological activity of compounds from natural products by means of half maximal inhibitory concentration (IC_{50}). Besides the determination of the parasitocidal effect itself, most studies in the literature also describe the cytotoxicity of natural products toward a mammalian cell line. Taking both measures into account, the *selectivity index* (SI) can be calculated by dividing the IC_{50} value of a compound for a mammalian cell line through the IC_{50} for its parasitocidal action. The SI indicates whether a compound is generally cytotoxic (low SI) or specifically affects the parasite (high SI). Compounds with high SI values are suitable for in vivo studies (Gehrig and Efferth 2008). The ratio of cytotoxicity to biological activity defined as selectivity index (SI) generally considers that biological efficacy is not due to the in vitro cytotoxicity when $SI \geq 10$ (Vonthron-Sénécheau et al. 2003). Some papers indicate the activity of a reference drug (standard) to allow the reader at least some comparability between different investigations.

Screening natural products provides the chance to discover new molecules of unique structure with high activity and selectivity which can be further optimized by semi- or fully synthetic procedures.

1.3 Natural Antimalarials

1.3.1 Antimalarials Isolated from Higher Plants

1.3.1.1 Antiplasmodial Activity of Plant Extracts

Turkish plants were evaluated for antiplasmodial activity in vitro against the multidrug resistant-K1 *Plasmodium falciparum* strain and for their ability to inhibit the purified enoyl-ACP reductase (FabI), a crucial enzyme in the fatty acid biosynthesis of *P. falciparum*. The most potent antiplasmodial activity was observed with the chloroform extracts of *Phlomis kurdica*, *P. leucophracta*, *Scrophularia cryptophila*, *Morina persica*, and the aqueous root extract of *Asperula nitida* subsp. *subcapitellata* (IC_{50} ranging from 1.5 to 1.9 µg/mL). The chloroform extract of *Rhododendron ungernii* leaves (IC$_{50}$ 10 μg/mL) and the water-soluble portion of *Rhododendron smirnovii* leaves (IC$_{50}$ 0.4 μg/mL) strongly inhibiting the FabI enzyme. The preliminary data indicate that some (poly) phenolic compounds are responsible for the FabI inhibition potential of these extracts (Tasdemir et al. 2005).

Ten plant extracts commonly used by the Meru community of Kenya were evaluated for the in vitro antiplasmodial, in vivo antimalarial, cytotoxicity, and animal toxicity activities. The water and methanol extracts of *Ludwigia erecta* and the methanol extracts of *Fuerstia africana* and *Schkuhria pinnata* exhibited high antiplasmodial activity (IC$_{50}$ < 5 μg/mL) against chloroquine sensitive (D6) and resistant (W2) *Plasmodium falciparum* clones. The cytotoxicity of these highly active extracts on Vero E6 cells were in the range 161.5–4650.0 μg/mL with a selectivity index (SI) of 124.2–3530.7. In vivo studies of these extracts showed less activity with chemosuppression of parasitaemia in *Plasmodium berghei* infected mice of 49.64–65.28 %. The methanol extract of *Clerodendrum erio-phyllum* with a lower in vitro activity (IC$_{50}$ 9.5–10.5 μg/mL) exhibited the highest chemosuppression of 90.1 %. The methanol and water extracts of *Pittosporum viridiflorum* were toxic to mice but at a lower dose prolonged survival of *P. berghei* infected mice with no overt signs of toxicity. However, the extracts were cytotoxic (SI, 0.96–2.51) on Vero E6 cells. Mostly, the water extracts showed weak activity or were inactive in vitro, but most of them were active in vivo, probably through biotransformation of chemical constituents into active metabo-lites. This may explain the use of these plants in traditional therapy (Muthaura et al. 2007).

1.3.1.2 Antiplasmodial Activity of Chemically Defined Molecules

Alkaloids

Among the natural products, indole alkaloids represent an interesting class of compounds. The review of Frederich et al. (2008) covers the indole alkaloids with high antiplasmodial activity (in vitro and in vivo) isolated from natural sources, and they are organized according to the different chemical structures of the alkaloids. Screening carried out till date has revealed several substances active in vitro under the micromolar range and with a selectivity index ranging from 0.9 to 375. Nevertheless, in vivo activity has been confirmed only in a small number of cases, and there is a need to undertake research focused on the mode of action of these compounds.

Terpenes

The possibility of developing new antimalarial drugs that could interfere with the biosynthesis of the dolichols, with the isoprenic chain of ubiquinones, and with protein isoprenylation led Lopes et al. (1999) to study the effects of different terpenes purified from essential oils. In the essential oil obtained from adult and plantlet leaves of *Viola surinamensis* 11 monoterpenes, 11 sesquiterpenes, and 3 phenylpropanoids were identified. Plantlet essential oil caused 100 % of growth inhibition after 48 h in the development of the young trophozoite to schizont stage and the sesquiterpene nerolidol (100 μg/ml) was identified as one of the active constituents (100 % of growth inhibition was obtained). In addition, examination of [U^{14}C]-glucose incorporation showed that activity of nerolidol is related to the inhibition of glycoprotein biosynthesis.

Goulart et al. (2004) investigated the effects of various terpenes (farnesol, nerolidol, limonene, and linalool) on the biosynthesis of dolichol and the isoprenic side chain of ubiquinones as well as on protein isoprenylation in the intraerythrocytic stages of *P. falciparum* and the 50 % inhibitory concentrations for each one were found: farnesol, 64 μM; nerolidol, 760 nM; limonene, 1.22 mM; linalool, 0.28 mM. The inhibitory effect of terpenes on the biosynthesis of both dolichol and the isoprenic side chain of ubiquinones and the isoprenylation of proteins in the intraerythrocytic stages of *P. falciparum* appears to be specific, because overall protein biosynthesis was not affected.

Quasinoids are a group of degraded triterpenes found in the family Simaroubaceae. Quassinoids with promising antiplasmodial activity were isolated from the roots of *Simaba orinocensis*. These quasinoids inhibited protein biosynthesis in vitro in translation system from the Krebs cycle cells. During intraerythrocytic proliferation, the malaria parasite makes its own ribosome, and the selective antimalarial action of these quassinoids may be explained by their stronger binding on the parasite ribosomes than binding to the host cells ribosomes (Muhammad et al. 2004).

Although quassinoids are generally cytotoxic, few compounds isolated from *Simmarrouba amara, Quassia amara,* and *Brucea javanica* were relatively selective against *P. falciparum* in vitro but were toxic in vivo. The antiplasmodial and cytotoxic properties of the quassinoids are possibly due to protein synthesis inhibition, and it is likely that parasite and host cell ribosomes are too similar to allow for the development of selective inhibitors (Wright 2005).

Flavonoids

Flavonoids are a ubiquitous group of compounds with a wide distribution in fruits and vegetables. De Monbrison et al. (2006) demonstrated in vitro antiplasmodial activity of flavonoid derivative dehydrosilybin on several strains of *P. falciparum*. The exact mechanism of antimalarial action of flavonoids is unclear, but some flavonoids are shown to inhibit the influx of L-glutamine and myoinositol into infected erythrocytes (Kaur et al. 2009).

Chalcones

Chalcones (1,3-diaryl-2-propen-1-ones) are compounds belonging to the flavonoid family and possess a broad spectrum of biological activities. Interest in the antimalarial activities of chalcones was prompted by the discovery of the antiplasmodial activity of licochalcone A, an oxygenated chalcone found in the roots of the Chinese licorice during routine screening. The last experiments demonstrate

that licochalcone A is a potent membrane-active agent, causing rapid and con-
centration-dependent transformation of discocytes into echinocytes paralleling the
antiplasmodial activity (Ziegler et al. 2004).

Fröhlich et al. (2005) analyzed the influence of the effect of prenylated chalcones of hops (*Humulus lupus*) on glutathione (GSH)-dependent hemin degradation to determine its contribution to the antimalarial effect of xantohumol and other
derivatives. The results demonstrate for the first time the ability of chalcone
derivatives to interfere with the hemin-degradation process of *P. falciparum*. This
effect might contribute to their antiplasmodial activity.

Quinones

Hydroxy-naphtoquinone isopinnatal possessing good antimalarial activity was
reported from *Kigelia pinnata*. The mode of action of this compound appears to be
the inhibition of mitochondrial electron transport in the respiratory chain by
reduced oxygen consumption similar to that of atovaquone (Weiss et al. 2000).

The structure of many naturally occurring quinones is based on the benzoqui-
none, naphthoquinone, or anthraquinone ring system. Naphthoquinones are rather
promising as blood schizonticides, since they are highly active against *P. falciparum*
in vitro. Roots of *Nephentes thorelii* yielded plumbagin and methyl 2-naphthazarin
both of which were evaluated against *P. falciparum* (Saxena et al. 2006).

Miscellaneous antimalarials from nature

The major surface protein of *P. falciparum* sporozoites, the circumsporozoite
protein (CSP) is proteolytically processed by a parasite-derived cysteine protease,
and this event is temporally associated with sporozoite invasion of host cells.
Allicin, a cysteine protease inhibitor found in garlic extracts was tested for its
ability to inhibit malaria infection (Coppi et al. 2006). It was shown that allicin,
present in freshly crushed garlic cloves, significantly inhibits sporozoite infectivity
in vivo and decreases parasite loads in mice with blood-stage infections. These
experiments demonstrate the feasibility of using the same cysteine protease
inhibitor to target two different life cycle stages in the vertebrate host and support
the idea that cysteine protease inhibitors may be useful drugs for the prophylaxis
and treatment of malaria.

Curcumin (a phenolic diketone) was potent against both chloroquine-sensitive
and -resistant *Plasmodium falciparum* strains (Cui et al. 2007). Curcumin's pro-
oxidant activity promoted the production in *P. falciparum* of reactive oxygen
species (ROS), whose cytotoxic effect could be antagonized by coincubation with
antioxidants and ROS scavengers. Curcumin treatment also resulted in damage of
both mitochondrial and nuclear DNA, probably due to the elevation of intracellular
ROS. Curcumin also inhibited the histone acetyltransferase (HAT) activity of the
recombinant *P. falciparum* general control non-derepressed 5 (PfGCN5) in vitro
and reduced nuclear HAT activity of the parasite in culture. Curcumin-induced
hypoacetylation of histone H3 at K9 and K14, but not H4 at K5, K8, K12, and
K16, suggested that curcumin caused specific inhibition of the PfGCN5 HAT.

These results indicated that curcumin exerts its antimalarial effect due to the generation of reactive oxygen species that damage both mitochondrial and nuclear DNA and downregulation of the PfGCN5 HAT activity.

Representative antiplasmodial drugs are presented in the Table 1.1.

A review of ethnopharmacological publications on antimalarial therapies from some Kenyan medicinal plants was presented by Muthaura et al. (2011). Compounds with remarkable antiplasmodial activities were isolated through bioassay-guided fractionation of active plant extracts and divided into chemical classes. Ideally, extracts and compounds effective at the blood stage of the malaria parasite should have strong in vitro antiplasmodial and in vivo antimalarial activities with good selectivity for the malaria parasites in cytotoxic assays. The review of Gelb (2007) summarizes some of the antimalarial drug discovery efforts.

1.3.2 Antiplasmodials Isolated from Bacteria, Fungi and Marine Organisms

The review of Fattorusso and Taglialatela-Scafati (2009) highlights the contribution of *marine chemistry* in the field of antimalarial research. About 60 secondary metabolites produced by marine organisms have been grouped into three structural types and discussed in terms of their reported antimalarial activities. The major groups of metabolites include isonitrile derivatives, alkaloids, and endoperoxide derivatives.

The most remarkable example of a marine compound with potential activity against *Plasmodium* is *manzamine* A, a unique beta-carboline alkaloid isolated from Indo-Pacific sponges (*Petrosiidae*), initially described as antitumor agent. More than 90 % of the asexual erythrocytic stages of *P. berghei* were inhibited after a single intraperitoneal injection of manzamine A into infected mice. Such suppressive activity is comparable to that of chloroquine and superior to that of artemisinin at the same dose. A remarkable aspect of manzamine A treatment is its ability to prolong the survival of highly parasitemic mice, with 40 % recovery 60 days after a single injection. Oral administration of an oil suspension of manzamine A also produced significant reductions in parasitemia (Ang et al. 2000).

Jasplakinolide is a cyclic peptide isolated from the marine sponge *Jaspis* sp., which markedly decreased parasitemia of *P. falciparum* by virtue of an apical protrusion that appears to interfere with the erythrocyte invasion by the merozoites and whose mechanism of formation is possibly related to an increase in F-actin content of the treated merozoites. The decrease became evident at day 2 at concentrations of 0.3 μM and above, and parasites finally disappeared at day 4 (Mizuno et al. 2002).

Recent research suggests that *marine organisms* may produce compounds with activity against malaria parasites. Three species of Caribbean tunicates were chosen, as more than 50 % reduction of *P. berghei* parasitemia was produced after administration of 250 or 500 mg/kg of their crude extracts into infected mice.

Table 1.1 Representative curative (C), preventive (P), and on-clinical-trial (T) antiplasmodial drugs (Laurent and Pietra 2006)

Chemical class or bioactivity	Drug	Mechanism of action	Efficacy
4-Aminoquinolines	C and P: chloroquine, amodiaquin		Safe; resistance detected
8-Aminoquinolines Antimicrobials	C: primaquine C: tetracycline, clindamycin, and fluoroquinolones; C and P: doxycycline	Inhibition of gametocytes Inhibition of protein biosynthesis in the apicoplast	
4-Carbinolquinolines Cinchona alkaloids	P: mefloquine C: quinine, quinidine	Ferriprotoporphyrin IX binding	Resistance detected
Endoperoxides	C: artemisinin, artemeter, arteether, artesunate; T: OZ277	C-radical formation with binding to proteins	
Iron chelating agents Naphthoquinones	C: desferoxamine C and P: atovaquone (administered in mixture with proguanil)	Inhibition of mitochondrial electron transport	
Phosphonates	T: fosmidomycin	Inhibition of DOXP reductoisomerase	
Pteridine analogs (Type-2 antifolates)	C: pyrimethamine; C and P: cycloguanil (metabolite of proguanil)	Inhibition of nucleic acid biosynthesis, targeting dihydrofolate reductase	Slow; resistance detected; better in combination with Type 1 antifolates
Sulfones, sulfonamides (Type-1 antifolates)	Dapsone, sulfadoxine, sulfadiazine, sulfalene	Inhibition of nucleic acid biosynthesis, targeting dihydropteroate synthase	Toxic or difficult to excrete; used by the parasite; resistance detected

The aqueous extracts of *Microcosmus goanus, Ascidia sydneiensis,* and *Phallusia nigra* were partitioned between water and n-butanol; the organic phases inhibited *P. falciparum* growth by 50 % at concentrations of 17.5 μg/ml, 20.9 μg/ml, and 29.4 μg/ml, respectively (Mendiola et al. 2006).

Laurent et al. (2006) screened for inhibitors of Pfnek-1, a protein kinase of *Plasmodium falciparum,* in south Pacific marine sponges. The ethanolic crude extract of a new species of *Xestospongia* was selected from which xestoquinone was isolated which inhibits Pfnek-1 with an IC_{50} around 1 μM. Among a small panel of plasmodial protein kinases, xestoquinone showed modest protein kinase inhibitory activity toward PfPK5 and no activity toward PfPK7 and PfGSK-3. Xestoquinone showed in vitro antiplasmodial activity against an FCB1 *P. falciparum* strain with an IC_{50} of 3 μM and a weak selectivity index (SI 7). Xestoquinone exhibited a weak in vivo activity at 5 mg/kg in *Plasmodium berghei* NK65 infected mice and was toxic at higher doses. Antiplasmodial marine natural products in the perspective of current chemotherapy and prevention of malaria is discussed in the review of Laurent and Pietra (2006).

1.4 Natural Antitrypanosomals

1.4.1 Antitrypanosomals Isolated from Higher Plants

1.4.1.1 Trypanocidal Activity of Plant Extracts

Extracts of different plant species growing in the Brazilian Cerrado were screened in vitro for trypanocidal activity. High-resolution gas chromatography analysis of the *n*-hexane extract of *T. stenocarpa* (IC_{50} = 23.6 μg/mL), the most active extract among all the tested samples, allowed the identification of β-amyrin, α-amyrin, lupeol, friedelin, β-friedelanol, campesterol, stigmasterol, and β-sitosterol. Oleanolic and ursolic acids were isolated from the methylene chloride extract of *T. stenocarpa* (IC_{50} = 51.5 μg/mL), while ursolic acid was isolated from the methylene chloride extract of *M. variabilis* (IC_{50} = 38.4 μg/mL). Solasonine and solamargine were identified as major compounds in the hydroalcoholic extract of the fruits of *S. lycocarpum* (IC_{50} = 57.1 μg/mL). The results showed that the trypanocidal activity may be related to the major compounds identified in the crude active extracts (Cunha et al. 2009).

The total extract (methanol) of the leaves of *Teclea trichocarpa* (Engl.) (Rutaceae), used traditionally in Kenya, yielded three acridone alkaloids, a furoquinoline alkaloid, and two triterpenoids. The isolated compounds were screened in vitro for cytotoxicity and against the parasitic protozoa, *Plasmodium falciparum, Trypanosoma brucei rhodesiense, Trypanosoma cruzi,* and *Leishmania donovani.* Among the compounds, α-amyrin had the greatest antiplasmodial activity (IC_{50} = 0.96 μg/ml), normelicopicine and skimmianine had the greatest antitrypanosomal activity against *T. b. rhodesiense* (IC_{50} = 5.24 μg/ml) and *T. cruzi*

(IC_{50} = 14.50 µg/ml), respectively. Normelicopicine also exhibited best antileishmanial activity (IC_{50} = 1.08 µg/ml). Arborinine exhibited moderate cytotoxicity (IC_{50} = 12.2 µg/ml) against L-6 cells. The compounds with low antiprotozoal and high cytotoxicity IC_{50} values are potential sources of template drug against parasitic protozoa (Mwangi et al. 2010).

Extracts of *Artemisia roxburghiana, Roylea cinerea, Leucas cephalotes, Nepeta hindostana, and Viola canescens* showed good antiplasmodial activity (IC_{50} < 5 µg/ml). The chloroform extract of *Artemisia roxburghiana* was the most active (IC_{50} value of 0.42 µg/ml) and the most selective (SI = 78) extract for *Plasmodium falciparum* among all plants extracts examined. The chloroform extract of *Leucas cephalotes* and the petroleum ether extract of *Viola canescens* exhibited substantial activities against *Leishmania donovani* with IC_{50} values of 3.61 µg/ml (SI = 8) and 0.40 µg/ml (SI = 30), respectively. The petroleum ether extract of *Viola canescens* exhibited activity against *Trypanosoma cruzi* with an IC_{50} value of 1.86 µg/ml (SI = 7). These results support investigation of components of traditional medicines as potential new antiprotozoal agents (Dua et al. 2011).

1.4.1.2 Trypanocidal Activity of Chemically Defined Molecules

Quinones

Bioactive quinones were isolated from the hartwood of trees of the Bignoniacae family (*Tabebuia spp.*). Among natural naphthoquinones, lapachol, β-lapachone, and its α-isomer have demonstrated trypanocidal activity. They can also be found in other families such as Verbenaceae, Proteaceae, Leguminosae, Sapotaceae, Scrophulariaceae, and Malvaceae. *β-Lapachone*, was found to be cytotoxic to a variety of human cancers and some of this mechanism of action is also responsible for its trypanocidal behavior. The trypanocidal action of these quinones is probably mediated by the production ROS and electrophilic metabolites, which bind to and inactivate *T. cruzi* macromolecules (Salas et al. 2008; Salas et al. 2011).

Juglone and plumbagin have been found to be active against epimastigote forms, leading to the total lysis of bloodstream trypomastigotes at a concentration similar to that of crystal violet, the standard drug recommended for the chemoprophylaxis of banked blood. The related diospyrin, a dimer of 7-methyljuglone isolated from the Indian plant *Diospyros montana*, and four synthetic derivatives were assayed on intracellular forms of *T. cruzi*. The dimethyl ether derivative was found to be more active than the parent compound (Hoet et al. 2004).

It is known that *naphthoquinones* are reduced by trypanothione reductase (TR), generating oxygen species through a redox cycle via the reaction of the reduced quinone with molecular oxygen. Enzymatic and antitrypanosomal studies have revealed that several naphthoquinone derivatives with strong trypanocidal activity were among the most effective TR inhibitors, suggesting that naphthoquinone reduction by parasitic flavoenzymes is a promising strategy for the development of new trypanocidal drugs (Blumenstiel et al. 1999).

Terpenes

Diterpene *5-epi-icetexone (ICTX)*, isolated from aerial parts of *Salvia gilliessi* exerts an antiproliferative effect on the epimastigotes of *Trypanosoma cruzi* from concentrations of 2.8 μM of the drug. At concentrations higher than 4.2 μM the drug became deleterious to the parasites with an IC_{50} 6.5 μM for 24 h exposure, and comparable to the effect of the trypanocidal drug benznidazole (IC_{50} 7.8 μM). ICTX was less cytotoxic for other cell types, such as cultured macrophages, because of the mortality that did not exceed 15 % at lethal dose for *T. cruzi*. It is possible that the deleterious effect was due to an oxidative action of the compound through the quinone groups. It is known that *T. cruzi* is highly sensitive to oxidative stress because the parasite has trypanothione–trypanothione reductase system in place of glutathione–glutathione reductase. Thus, they maintain their reduced glutathione by a non-enzymatic reaction, which is not defense enough against peroxidative damage. The cytostatic effect of ICTX may be due to an interaction with the DNA of the parasite, as occurs with other natural compounds. Moreover, an inhibitory effect of ICTX on *taq*-DNA polymerase has been confirmed and it could cause an arrest at the S phase of the cell cycle (Sanchez et al. 2006).

Helenalin and mexicanin I are sesquiterpene lactones from *Arnica* and *Inula* species (Asteraceae). The IC_{50} values on bloodstream forms of *T. b. rhodesiense* were 0.051 and 0.318 μM, respectively, and SI values of 19.5 and 7.7 were obtained when compared to L6 cells. For the reference drug, melarsoprol, an IC_{50} value of 0.008 μM was observed. It has been suggested that the activity of the compounds is due to the presence of an $α,β$-unsaturated carbonyl group, which alkylates biological nucleophils. Moreover, the compounds display two alkylation centers, a cyclo-pentenone and the $α$-methylene-$γ$-lactone moieties, mediating interactions with sulphydryl groups of various enzymes. This bifunctionality is hypothesized to increase the specificity toward the parasite compared to monofunctional sesquiter-pene lactones, which possessed higher cytotoxicity than trypanocidal activity. In addition, stereochemistry seems to play an important role, since mexicanin I, a diastereoisomer of helenalin, was six times less active. The mode of action is anticipated to involve the trypanothione metabolism to increase oxidative stress in the parasite (Schmidt et al. 2002; Gehrig and Efferth 2008).

Mishina et al. (2007) have found that *artemisinin and its derivatives* are effective against *T. cruzi* and *T. brucei rhodensisnse*. Artemisinin inhibited 50 % of growth (IC_{50}) at low micromolar concentrations, 13.4 μM for *T. cruzi* and 20.4 μM for *T. brucei rhodensiense*. The IC_{50} values for the trypanosomes are comparable to those obtained with *L. donovani* promastigotes (30.8 μM). Arte-misinin inhibits calcium-dependent ATPase activity in *T. cruzi* membranes, sug-gesting a mode of action via membrane pumps.

Flavonoids

Tasdemir et al. (2006) screened a large number of flavonoids and analogs for their trypanocidal activity in vitro. The most potent compound was the natural product 7,8-dihydroxyflavone, followed by 3-hydroxyflavone, rhamnetin, and 7,8,3',4'-tetrahydroxyflavone with IC_{50} values of 0.068, 0.5, 0.5 and 0.5 μg/ml, respectively, on bloodstream forms of *T. b. rhodesiense* and SI values of 116.2, 28.8, >180, and 43.2 when compared to L6 cells. 7,8-Dihydroxyflavone and quercetin appeared to ameliorate parasitic infections in mouse models. Generally, the test compounds lacked cytotoxicity in vitro and in vivo.

Several natural compounds including alkaloids, phenolic derivatives, quinones, and terpenes have been shown to inhibit the growth of trypanosomes in vitro with GI_{50} (growth inhibition) values in the submicromolar range (Hoet et al. 2004). One interesting lead compound is ascofuranone, a prenylated phenol antibiotic produced by the phytopathogenic fungus *Ascochyta visiae*. Ascofuranone is an inhibitor of the trypanosome alternative oxidase, a unique mitochondrial electron transport system in these parasites. A recent study has shown that ascufuranone can cure *T. brucei*infected mice if the compound is administered intraperitoneally for four consecutive days at 100 mg/kg or orally for eight consecutive days at 400 mg/kg (Yabu et al. 2003).

Over the past decade, several new lead compounds for the treatment of human trypanosomiases have been identified. Inhibitors of ergosterol synthesis are encouraging anti-*T. cruzi* compounds as they target an essential metabolic pathway of this parasite. Peptidyl and non-peptidyl cysteine protease inhibitors are very promising antitrypanosomal agents as these compounds are small in size and inexpensive to produce. Other promising compounds are the DNA topoisomerase and proteasome inhibitors currently used in cancer chemotherapy (Steverding and Tyler 2005).

1.4.2 Antitrypanosomals Isolated from Fungi, Bacteria and Marine Organisms

Alkaloids

Ascididemin and 2-bromoascididemin, pyridoacridone alkaloids derived from a marine organism (*Cystodytes* sp., Ascidiacea) possessed trypanocidal activity in vitro. Ascididemin is known to exhibit antitumor activity by DNA intercalation. The IC_{50} values were 4 and 1.76 μg/ml (14 and 5 μM), respectively, on bloodstream forms of *T. b. rhodesiense*, and SI values of 0.1 and 0.4 were observed when compared to L6 cells. For the reference drug melarsoprol an IC_{50} value of 0.0012 μg/ml (0.003 μM) was obtained. The presence of a benzonaphthyridone substructure seems to be important for the trypanocidal activity. It has been suggested that the compounds intercalate into the DNA and induce oxidative damage. Nevertheless, the potential of the substances is limited due to their high cytotoxicity. By synthetic derivatization related compounds were obtained, which possessed significantly higher specificity toward the parasite—the IC_{50} values were 0.002 and 0.005 μg/ml on bloodstream forms of *T. b. rhodesiense*, and SI values 950 and 316 were observed when compared to L6 cells (Copp et al. 2003).

Chlorinated aromatic compounds

Ambigol A and ambigol C, two chlorinated aromatic compounds, were isolated from the terrestrial cyanobacterium *Fischerella ambigua*. The minimal inhibitory concentrations (MIC) were 33 and 11 µg/ml, respectively, on bloodstream forms of *T. b. rhodesiense*. Whereas ambigol A showed no selectivity toward the parasite with an SI value of 1 when compared to L6 cells, the SI value for ambigol C was 9.1 (Wright et al. 2005).

1.5 Natural Antileishmanials

1.5.1 Antileishmanials Isolated from Higher Plants

1.5.1.1 Leishmanicidal Activity of Plant Extracts

The oil of *Croton cajucara* has been used successfully against *Leishmania amazonensis*. Morphological changes in *L. amazonensis* promastigotes were observed in vitro within 1 h following the application of 15 ng/ml of oil, leading to nuclear and kinetoplast chromatin destruction followed by cell lysis. Treatment of preinfected murine macrophages with 15 ng/ml of oil caused a 50 % reduction in *L. amazonensis* promastigotes infecting macrophages and a 60 % increase in macrophage NO production in preinfected macrophages (Rosa et al. 2003).

The dichloromethanic and ethanolic crude *extracts* (flowers and leaves) and flavonoids (including luteolin and quercetin) isolated from *Chromolaena hirsuta* (Asteraceae) have been assayed for antiprotozoal activity against trypomastigote forms of *Trypanosoma cruzi* and promastigote forms of *Leishmania amazonensis*. The crude extracts significantly reduced the viability of *T. cruzi amazonensis* (20.27–30.72 % of viable trypomastigotes) and *L. amazonensis*-promastigotes (30.20–86.42 % of viable promastigotes). The flavonoids clearly reduced the viability of *T. cruzi*-trypomastigotes with IC_{50} between 102.40 and 352.60 µg/ml (Mittra et al. 2000; Taleb-Contini et al. 2004).

The ethanolic *extract* from *Yucca filamentosa* L. showed the strongest leishmanicidal activity (100 % inhibition at 5 µg/ml). The bioactivity-guided fractionation of this extract led to the isolation of three main components (yuccasaponins MC 1–3). The effect of yuccasaponin MC 3 on the promastigote form of *L. mexicana amazonensis* was quantified and characterized using flow cytometry and specific fluorescent dyes. The data revealed that the membrane of the promastigote is attacked. By this method, an inhibition of intracellular growth of *L. major* was demonstrated (Plock et al. 2001).

Ayurveda or ayurvedic medicine is a Hindu system of traditional medicine native to India and a form of alternative medicine. The earliest literature on Indian medical practice appeared during the Vedic period in India, i.e., in the mid-second millennium BCE. Ayurveda stresses the use of plant-based medicines and treatments. Byadgi (2011) presented a review of plants used in the Ayurvedic system of medicine with an aim to use them for the treatment of various diseases. The antileishmanial activity of arbortristoside A (isolated from the plant *Nycanthensis arbortristitis*) has been confirmed in hamsters. Antileishamanial activity in vivo showed crude extracts of the following plants: *Tephrosia purpurea, Amoora rohituka, Swertia chirata, Tibouchina semidecandra, Tinospora cordifolia.*

1.5.1.2 Leishmanicidal Activity of Chemically Defined Molecules

Flavonoids

Luteolin and quercetin isolated from *Vitex negundo* (Verbenaceae) and *Fagopyrum esculentum* (Polygonaceae) are potent antileishmanial compounds with IC_{50} values against *L. donovani* intracellular amastigotes of 12.5 and 45.5 μM, respectively. Both compounds are able to induce topoisomerase II-mediated kinetoplastid DNA minicircle cleavage in *L. donovani* promastigotes and intracellular amastigotes. Treatment of promastigotes with luteolin and quercetin leads to cell cycle arrest in the G0/G1 phase followed by apoptotic cell death. In vivo studies have shown that luteolin reduced the splenic parasite load of infected rodents by 80 % (3.5 mg/kg body weight) and quercetin by 90 % (14 mg/kg body weight). Luteolin appears to be non-toxic to normal human T-cells, though quercetin induced cell cycle arrest. It has also been reported that luteolin and quercetin are specific inhibitors of topoisomerase I, which is an unusual bi-subunit topoisomerase in *Leishmania* (Das et al. 2006).

Chalcones

Licochalcone A, an oxygenated chalcone isolated from Chinese licorice *Glycyrrhiza spp.* (Fabaceae), exhibits strong antileishmanial activity, markedly preventing the growth of *L. major* and *L. donovani* promastigotes and amastigotes. IC_{50} value of licochalcone A against *L. donovani* intracellular amastigote was 0.9 μg/ml (2.7 μM) and 7.2 μg/ml (21 μM) against *L. major* promastigotes. In vivo studies with hamsters have shown that the parasite load in the spleen and liver was reduced up to 96 %. Other oxygenated chalcones exhibited comparable potency. No cytotoxic effects have been observed in human leucocytes, lymphocytes, and monocytes. Licochalcone A and related chalcones are able to destroy the mitochondrial ultrastructure. Furthermore, the compounds are strong inhibitors of fumarate reductase in *L. major* (Chen et al. 2001).

Saponins

Saponins α-hederin, s-hederin, and hederacolchiside A1 derived from *Hedera helix* (Araliaceae) exhibited antileishmanial properties. A1 appeared to be the most prominent compound with an IC_{50} value of 1.2 μM against *L. infantum* promastigotes and 0.053 μM against intracellular amastigotes. Moderate cytotoxicity was observed in human monocytes (Delmas et al. 2000). Moreover, it has been shown that α-hederin is able to induce nitric oxide production in murine macrophages (Jeong and Choi 2002).

Lignans

Diphyllin, isolated from *Haplophyllum bucharicum* (Rutaceae), showed antileishmanial activity against *L. infantum* promastigotes and intracellular amastigotes. IC_{50} values were 14.4 µg/ml (38 µM) and 0.2 µg/ml (0.5 µM), respectively. The lower IC_{50} value for amastigotes results from inhibition of parasite uptake within macrophages, which is probably due to surface molecule modifications. Diphyllin has been shown to be anti-proliferative against human monocytes [IC_{50} value 35.2 µg/ml (93 µM)] (Di Giorgio et al. 2005).

Monoterpenes

Linalool, a monoterpene extracted from *Croton cajucara* (Euphorbiaceae), exhibits strong antileishmanial activity against *L. amazonensis* promastigotes and intracellular amastigotes. For purified linalool, the LD_{50} values were 4.3 ng/ml (28 nM) and 22 ng/ml (143 nM). Treatment of preinfected murine macrophages with 15 ng/ml of linalool-rich essential oil reduced the interaction between the macrophages and the parasite by 50 %, associated with an increased nitric oxide production. Treatment for 1 h destroyed 100 % of both promastigotes and intracellular amastigotes, while exhibiting no cytotoxic effect to murine macrophages. In vitro, chromatin plus kinetoplastid destruction and mitochondrial swelling have been observed, followed by cell lysis (Do Socorro et al. 2003).

Sesquiterpenes

The discovery of artemisinin as a pharmaceutical for the treatment of malaria promoted an interest in the discovery of new compounds from plants with antiprotozoal activity, especially those to treat diseases caused by *Leishmania* parsites (Chan-Bacab and Peña-Rodríguez 2001). Dihydroartemisinin reduces parasite burdens by 75 % in the spleens and livers of hamsters infected with *Leishmania donovani* (Ma et al. 2004).

Nerolidol, a sesquiterpene present in the essential oils of some plants—neroli, ginger, jasmine, lavender, tea tree, and lemongrass, inhibited the growth of *Leishmania amazonensis*, *L. braziliensis*, and *L. chagasi* promastigotes, and *L. amazonensis* amastigotes with in vitro 50 % inhibitory concentrations of 85, 74, 75, and 67 µM, respectively. The treatment of *L. amazonensis*-infected macrophages with 100 µM nerolidol resulted in 95 % reduction in infection rates. This compound effect can be attributed to the blockage of an early step in ergosterol synthesis (Arruda et al. 2005).

Miscellaneous natural products

Coumarin, isolated from *Calophyllum brasiliense* (Clusiaceae), exhibits in vitro leishmanicidal effects on promastigote and amastigote forms of *L. amazonensis*. IC_{50} values were 3.0 µg/ml (7.4 µM) and 0.88 µg/ml (2.2 µM), respectively. This compound showed no cytotoxicity in murine macrophages at concentrations

Table 1.2 Natural products that showed antileishmanial activity (Monzote 2009)

Compound	Natural source	Antileishmanial activity
2',6'-dihydroxy-4'-methoxychalcone	*Piper aduncum*	Exhibited in vitro activity against promastigotes and amastigotes of *L. amazonensis*
Canthin-6-one alkaloids	*Zanthoxylum chiloperone*	Demonstrated in vivo activity in BALB/c mice infected with *L. amazonensis*
Coronaridine	*Peschiera australis*	Showed in vitro activity against promastigotes and amastigotes of *L. amazonensis*
Licochalcone A	*Chinese licorice*	Exhibited activity in vitro and in vivo against *L. major* and *L. donovani*
Maesabalide III	*Maesa balansae*	Caused in vitro and in vivo activity against *L. donovani*
Parthenolide	*Tanacetum parthenium*	Displayed activity against promastigotes and amastigotes of *L. amazonensis*
Plumbagin	*Pera benensis*	Demonstrated in vivo activity in BALB/c mice infected with *L. amazonensis* and *L. venezuelensis*
Trichothecenes	*Holarrhena floribunda*	Exhibited antileishmanial activity against promastigotes and amastigotes of *L. donovani*

Table 1.3 Potential new antileishmanial agents extracted from plants (Carvalho and Ferreira 2001)

Plant	Chemical class	Compound
Galipea longiflora	Quinoline alkaloids	Chimamine B
Ampelocera edentula	Tetralones	4-Hydroxyy-1-tetralone
Pera benensis	Naphthoquinones	8,8'- Biplumbagin
Rollinia emarginata	Acetogenin	Rolliniastatin-1, squamocin
Asparagus africanus	Lignan	(+)-Nyasol
Dictyoloma peruviana	4-Quinoline alkaloids	Dictyolomide A and B
Unonopsis buchtienii	Aporphine alkaloids	Unonopsine
Acanthus illicifolius	Oxazolinones	2-Benzoxazolinone
Picrorhiza kurroa	Irridoid glycosides	Picroliv

mentioned above. Treatment with the compound revealed ultra-structural changes in promastigotes, such as mitochondrial swelling and intense exocytotic activity at the flagellar pocket (Brenzan et al. 2007).

The most promising products with antileishmanial activity, isolated from natural sources, are summarized in Table 1.2.

Compounds, isolated from plants with promising activity against the *Leishmania* genera and low toxicity as compared to the antimonial drugs, are summarized in Table 1.3.

The most important advances obtained in study of antileishmanial activity of natural products during the recent years have been reviewed by Rocha et al. (2005), Salem and Werbovetz (2006) and Tempone et al. (2008). A brief of natural products that are promising leads for the development of novel leishmanicidal chemotherapeutics has been presented by Polonio and Efferth (2008). Perspectives of the further research into plant active components as a resource for antiparasitic agents are outlined in the review of Anthony et al. (2005).

1.5.2 Antileishmanials Isolated from Fungi, Bacteria and Marine Organisms

Some studies revealed the search of new products in microorganism or marine sources, such as a glycoprotein isolated from the sponge *Pachymatisma johnstonii*, which showed a high activity in vitro against *L. donovani, L. braziliensis,* and *L. mexicana* (Le Pape et al. 2000), and aphidicolin a fungal metabolite isolated from *Nigrospora sphaerica*, which inhibited the growth of promastigotes and amastigotes of *L. donovani* (Kayser et al. 2001).

1.6 Conclusion

According to many authors, such as Polonio and Efferth (2008), Monzote (2009) and Anthony et al. (2005), although a significant number of antimalarial, anti-trypanosomal, and anti-leishmanial evaluations, as well as cytotoxicity studies in vitro of natural compounds have been determined, the number of mechanistic studies seems to be rather small. It is necessary to search for actual target sites because they are unknown in most cases. Genomic and proteomic approaches should be used to investigate gene products that play a role in the mode of action of isolated natural substances. From a chemical point of view, derivatization of identified lead structures and evaluation of essential binding structures will contribute to the improvement of efficacy and specificity toward the parasite. It is important to validate therapeutic concepts and to evaluate toxicity of selected compounds in animal studies. This multidisciplinary approach is necessary in the further search for effective antiparasitic drugs isolated from natural sources.

References

Ang KKH, Holmes MJ, Higa T, Hamann MT, Kara UAK (2000) In vivo antimalarial activity of the beta-carboline alkaloid manzamine A. Antimicrob Agents Chemother 44(6):1645–1649. doi: 10.1128/AAC.44.6.1645-1649.2000

Anthony JP, Fyfe L, Smith H (2005) Plant active components—a resource for antiparasitic agents? Trends Parasitol 21:462–468. doi:10.1016/j.pt.2005.08.004

Arruda DC, D'Alexandri FL, Katzin AM, Uliana SRB (2005) Antileishmanial activity of the terpene nerolidol. Antimicrob Agents Chemother 48:1679–1687. doi: 10.1128/AAC.49.5.1679-1687.2005

Barret MP, Vincent IM, Burchmore RJ, Kazibwe AJ, Matovu E (2011) Drug resistance in human African trypanosomiasis. Future Microbiol 6(9):1037–1047. doi:10.2217/fmb.11.88

Blumenstiel K, Schoneck R, Yardley V, Croft SL, Krauth-Siegel RL (1999) Nitrofuran drugs as common subversive substrates of *Trypanosoma cruzi* lipoamide dehydrogenase and trypanothione reductase. Biochem Pharmacol 58:1791–1799

Brenzan MA, Nakamura CV, Prado Dias Filho B et al (2007) Antileishmanial activity of crude
    extract and coumarin from *Calophyllum brasiliense* leaves against *Leishmania amazonensis*.
    Parasitol Res 101:715–722. doi: 10.1007/s00436-007-0542-7
Burleigh BA, Woolsey AM (2002) Cell signaling and *Trypanosoma cruzi* invasion. Cell
    Microbiol 4:701–711
Byadgi PS (2011) Natural products and their antileishmanial activity. A critical review. Int Res J
    Pharm 2:46–49. http://www.irjponline.com
Carvalho PB, Ferreira EI (2001) Leishmaniasis phytotherapy. Nature's leadership against an
    ancient disease—review. Fitoterapia 72:599–618
Chan-Bacab MJ, Peña-Rodríguez LM (2001) Plant natural products with leishmanicidal activity.
    Nat Prod Rep 18:674–688. doi:10.1039/B100455G
Chawla B, Madhubala R (2010) Drug targets in *Leishmania*. J Parasit Dis 34:1–13. doi:10.1007/
    s12639-010-0006-3
Chen M, Zhai L, Christensen SB, Theander TG, Kharazmi A (2001) Inhibition of fumarate
    reductase in *Leishmania major* and *L. donovani* by chalcones. Antimicrob Agents Chemother
    45:2023–2029. doi:10.1128/AAC.45.7.2023-2029.2001
Copp RB, Kayser O, Brun R, Kiderlen AF (2003) Antiparasitic activity of marine pyridoacridone
    alkaloids related to the ascididemins. Planta Med 69:527–531. doi:10.1055/s-2003-40640
Coppi A, Cabinian M, Mirelman D, Sinnis P (2006) Antimalarial activity of allicin, a biologically
    active compound from garlic cloves. Antimicrob Agents Chemother 50:1731–1737. doi:10.
    1128/AAC.50.5.1731-1737.2006
Croft SL, Sundar S, Fairlamb AH (2006) Drug resistance in leishmaniasis. Clin Microbiol Rev
    19(1):111–126. doi: 10.1128/CMR.19.1.111–126.2006
Cui L, Miao J, Cui L (2007) Cytotoxic effect of curcumin on malaria parasite *Plasmodium
    falciparum*: inhibition of histone acetylation and generation of reactive oxygen species.
    Antimicrob Agents Chemother 51(2):488–494. doi:10.1128/AAC.01238-06
Cunha WR, dos Santos FM, Peixoto JA, Veneziani RCS, Crotti AEM, Silva MLA, Filho AAS,
    Albuquerque S, Turatti ICC, Bastos JK (2009) Screening of plant extracts from the Brazilian
    Cerrado for their in vitro trypanocidal activity. Pharm Biol (formerly Int J Pharmacognosy)
    47:744–749. doi: http://dx.doi.org/10.1080/13880200902951361
Das A, Dasgupta A, Sengupta T, Majumder HK (2004) Topoisomerases of kinetoplastid parasites
    as potential chemotherapeutic targets. Trends Parasitol 20:381–387. doi:10.1016/j.pt.2004.
    06.005
Das BB, Sen N, Roy A, Dasgupta SB, Ganguly A, Mohanta BC, Dinda B, Majumder HK (2006)
    Differential induction of *Leishmania donovani* bi-subunit topoisomerase I-DNA cleavage
    complex by selected flavones and camptothecin: activity of flavones against camptothecin-
    resistant topoisomerase I. Nucleic Acids Res 34:1121–1132. doi:10.1093/nar/gkj502
de Carvalho PB, Ferreira EI (2001) Leishmaniasis phytotherapy. Nature's leadership against an
    ancient disease. Fitoterapia 72:599–618
de Monbrison F, Maitrejean M, Latour Ch, Bugnazet F, Peyron F, Barron D, Staphane P (2006)
    In vitro antimalarial activity of flavonoid derivatives dehydrosilybin and 8-(1,1)-DMA-
    kaempferide. Acta Tropica 97:102–107
Delmas F, Di Giorgio C, Elias R et al (2000) Antileishmanial activity of three saponins isolated
    from ivy, alpha-hederin, beta-hederin and hederacolchiside A1, as compared to their action on
    mammalian cells cultured in vitro. Planta Med 66:343–347. doi:10.1055/s-2000-8541
Di Giorgio C, Delmas F, Akhmedjanova V et al (2005) In vitro antileishmanial activity of
    diphyllin isolated from *Haplophyllum bucharicum*. Planta Med 71:366–369
Do Socorro SRMS, Mendonca-Filho RR, Bizzo HR et al (2003) Antileishmanial activity of a
    linalool-rich essential oil from *Croton cajucara*. Antimicrob Agents Chemother 47:
    1895–1901. doi: 10.1128/AAC.47.6.1895-1901.2003
Dua VK, Verma G, Agarwal DD, Kaiser M, Brun R (2011) Antiprotozoal activities of traditional
    medicinal plants from the Garhwal region of North West Himalaya, India. J Ethnopharmacol
    136:123–128. doi: 10.1016/j.jep.2011.04.024

EAC (2012) http://www.eac.int/health/index.php?option=com_content&view=article&id=94%3 Acl

Eckstein-Ludwig U, Webb RJ, Van Goethem ID, East JM, Lee AG, Kimura M, O'Neill PM, Bray PG, Ward SA, Krishna S (2003) Artemisinins target the SERCA of *Plasmodium falciparum*. Nature 424:957–961

Fattorusso E, Taglialatela-Scafati O (2009) Marine antimalarials. Mar Drugs 7:130–152. doi:10.3390/md7020130

Frederich M, Tits M, Angenot L (2008) Potential antimalarial activity of indole alkaloids. Trans R Soc Trop Med Hyg 102:11–19. doi:10.1016/j.trstmh.2007.10.002

Fröhlich S, Schubert C, Bienzle U, Jenett-Siems K (2005) In vitro antiplasmodial activity of prenylated chalcone derivatives of hops (*Humulus lupulus*) and their interaction with haemin. J Antimicrob Chemother 55:883–887. doi:10.1093/jac/dki099

Gehrig S, Efferth T (2008) Development of drug resistance in *Trypanosoma brucei rhodensiense* and *Trypanosoma brucei gambiense*. Treatment of human African trypanosomiasis with natural products (Review). Int J Mol Med 22:411–419. doi:0.3892/ijmm_00000037

Gelb MH (2007) Drug discovery for malaria: a very challenging and timely endeavor. Curr Opin Chem Biol 11:440–445. doi:10.1016/j.cbpa.2007.05.038

Goulart HR, Kimura EA, Peres VJ, Couto AS, Duarte FAA, Katzin AM (2004) Terpenes arrest parasite development and inhibit biosynthesis of isoprenoids in *Plasmodium falciparum*. Antimicrob Agents Chemother 48:2502–2509. doi: 10.1128/AAC.48.7.2502–2509.2004

Hoet S, Opperdoes F, Brun R, Quetin-Leclercq J (2004) Natural products active against African trypanosomes: a step towards new drugs. Nat Prod Rep 21:353–364

Jeong HG, Choi CY (2002) Expression of inducible nitric oxide synthase by alpha-hederin in macrophages. Planta Med 68:392–396

Kaur K, Jain M, Kaur T, Jain R (2009): Antimalarials from nature. Bioorg Med Chem 17:3229–3256

Kayser O, Kiderlen AF, Bertels S, Siems K (2001) Antileishmanial activities of aphidicolin and its semisynthetic derivatives. Antimicrob Agents Chemother 45:288–292. doi:10.1128/AAC.45.1.288-292.2001

Kayser O, Kiderlen AF, Croft SL (2002) Natural products as potential antiparasitic drugs. In Studies in Natural Product Chemistry 26, pp 779–848. Elsevier, UK. http://userpage.fu-berlin.de/~kayser/antiparasiticsfromnature.pdf

Kayser O, Kiderlen AF, Croft SL (2003) Natural products as antiparasitic drugs. Parasitol Res 90:S55–S62. doi:10.1007/s00436-002-0768-3

Kothari H, Kumar P, Sundar S, Singh N (2007) Possibility of membrane modification as a mechanism of antimony resistance in *Leishmania donovani*. Parasitol Int 56(1):77–80. http://dx.doi.org/10.1016/j.parint.2006.10.005

Laport MS, Santos OCS, Muricy G (2009) Marine sponges: potential sources of new antimicrobial drugs. Curr Pharm Biotechnol 10:86–105

Laurent D, Pietra F (2006) Antiplasmodial marine natural products in the perspective of current chemotherapy and prevention of malaria. A review. Mar Biotechnol 8:433–447. doi:10.1007/s10126-006-6100-y

Laurent D, Jullian V, Parenty A, Knibiehler M, Dorin D, Schmitt S, Lozach O, Lebouvier N, Frostin M, Alby F, Maurel S, Doerig C, Meijerf M, Sauvain M (2006) Antimalarial potential of xestoquinone, a protein kinase inhibitor isolated from a Vanuatu marine sponge *Xestospongia* sp. Bioorg Med Chem 14:4477–4482. doi:10.1016/j.bmc.2006.02.026

Le Pape P, Zidane M, Abdala H, Moré MT (2000) A glycoprotein isolated from the sponge, *Pachymatisma johnstonii*, has anti-leishmanial activity. Cell Biol Int 24:51–56. doi:10.1006/cbir.1999.0450

Lopes NP, Kato MJ, Andrade EH, Maia JG, Yoshida M, Planchart AR, Katzin AM (1999) Antimalarial use of volatile oil from leaves of *Virola surinamensis* (Rol.) Warb. by Waiapi Amazon Indians. J Ethnopharmacol 67:313–319. http://dx.doi.org/10.1016/S0378-8741(99)00072-0

Ma Y, Lu DM, Lu XJ, Lia L, Hu XS (2004) Activity of dihydroartemisinin against *Leishmania donovani* both in vitro and in vivo. Chin Med J 117:1271–1273

Malebo HM, Tanja W, Cal M, Swaleh SAM, Omolo MO, Hassanali A, Séquin U, Hamburger M, Brun R, Ndiege IO (2009) Antiplasmodial, anti-trypanosomal, anti-leishmanial and cytotoxicity activity of selected Tanzanian medicinal plants. Tanzan J Health Res 11(4): 2266–2234

Maltezou HC (2010) Drug resistance in visceral leishmaniasis. J Biomed Biotechnol 2010:617521. doi: 10.1155/2010/617521

Maya JD, Cassels BK, Iturriaga-Vásquez P, Ferreira J, Faúndez M, Galanti N, Ferreira A, Morello A (2007) Mode of action of natural and synthetic drugs against *Trypanosoma cruzi* and their interaction with the mammalian host. Comp Biochem Physiol Part A 146:601–620. doi:10.1016/j.cbpa.2006.03.004

Mendiola J, Hernández H, Sariego I, Rojas L, Otero A, Ramírez A, Chávez MA, Payrol JA (2006) Antimalarial activity from three ascidians: an exploration of different marine invertebrate phyla. Trans R Soc Trop Med Hyg 100:909–915. doi:10.1016/j.trstmh.2005.11.013

Mishina YV, Krishna S, Haynes RK, Meade JC (2007) Artemisinins inhibit *Trypanosoma cruzi* and *Trypanosoma brucei rhodesiense* in vitro growth. Animicrob Agents Chemother 51(5):1852–1854. doi:10.1128/AAC.01544-06

Mittal MK, Rai S, Ravinger A, Gupta S, Sundar S, Goyal N (2007) Characterization of natural antimony resistance in *Leishmania donovani* isolates. Am J Trop Med Hyg 76(4):681–688

Mittra B, Saha A, Chowdhury AR et al (2000) Luteolin, an abundant dietary component is a potent anti-leishmanial agent that acts by inducing topoisomerase II-mediated kinetoplast DNA cleavage leading to apoptosis. Mol Med 6:527–541

Mizuno Y, Makioka A, Kawazu S, Kano S, Kawai S, Akaki M, Aikawa M, Ohtomo H (2002) Effect of jasplakinolide on the growth, invasion, and actin cytoskeleton of *Plasmodium falciparum*. Parasitol Res 88:844–848. doi:10.1007/s00436-002-0666-8

Monzote L (2009) Current treatment of leishmaniasis: a review. Open Antimicrob Agents J 1: 9–19

Muhammad I, Bedir E, Khan SI, Tekwani BL, Khan IA, Takamatsu S, Pelletier J, Walker LA (2004) A new antimalarial quassinoid from *Simaba orinocensis*. J Nat Prod 67:772–777

Mukherjee A, Padmanabhan PK, Singh S et al (2007) Role of ABC transporter MRPA, γ-glutamylcysteine synthetase and ornithine decarboxylase in natural antimony-resistant isolates of *Leishmania donovani*. J Antimicrob Chemother 59(2):204–211. doi:10.1093/jac/dkl494

Muthaura CN, Rukunga GM, Chhabra SC, Omar SA, Guantai AN, Gathirwa JW, Tolo FM, Mwitari PG, Keter LK, Kirira PG, Kimani CW, Mungai GM, Njagi EN (2007) Antimalarial activity of some plants traditionally used in Meru district of Kenya. Phytother Res 21:860–867. doi:10.1002/ptr.2170

Muthaura CN, Keriko JM, Derese S, Yenesew A, Rukunga GM (2011) Investigation of some medicinal plants traditionally used for the treatment of malaria in Kenya as potential sources of antmalarial drugs. Exp Parasitol 127:609–626. doi:10.1016/j.exppara.2010.11.004

Mwangi ESK, Keriko JM, Machocho AK, Wanyonyi AW, Malebo HM, Chhabra SC, Tarus PK (2010) Antiprotozoal activity and cytotoxicity of metabolites from leaves of *Teclea trichocarpa*. J Med Plants Res 4(9):726–731. doi:10.5897/JMPR09.188

Plock A, Sokolowska-Köhler W, Presber W (2001) Application of flow cytometry and microscopical methods to characterize the effect of herbal drugs on *Leishmania* spp. Exp Parasitol 97:141–153. doi:10.1006/expr.2001

Polonio T, Efferth T (2008) Leishmaniasis: drug resistance and natural products (Review). Int J Mol Med 22:277–286

Rocha LG, Almeida JRGS, Macedo RO, Barbosa-Filho JM (2005) A review of natural products with antileishmanial activity. Phytomedicine 12:514–535. doi:10.1016/j.phymed.2003.10

Rosa MSS, Mendonça-Filho RR, Bizzo HR, de Almeida RI, Soares RM, Souto-Padrón T, Alviano CS, Lopes AH (2003) Antileishmanial activity of linalool-rich essential oil from *Croton cajucara*. Antimicrob Agents Chemother 47:1895–1901. doi:10.1128/AAC.47.6.1895-1901.2003

Salas C, Tapia RA, Ciudad K, Armstrong V, Orellana M, Kemmerling U, Ferreira J, Maya JD, Morello A (2008) *Trypanosoma cruzi*: activities of lapachol and alpha- and beta-lapachone derivatives against epimastigote and trypomastigote forms. Bioorg Med Chem 16:668–674

Salas CO, Faúndez M, Morello A, Maya JD, Tapia RA (2011) Natural and synthetic naphthoquinones active against *Trypanosoma Cruzi*: an initial step towards new drugs for Chagas disease. Curr Med Chem 18:144–161

Salem MM, Werbovetz KA (2006) Natural products from plants as drug candidates and lead compounds against leishmaniasis and trypanosomias. Curr Med Chem 13:2571–2598

Sanchez AM, Jimenez-Ortiz V, Sartor T, Tonn CE, García EE, Nieto M, Burgos MH, Sosa MA (2006) A novel icetexane diterpene, 5-epi-icetexone from *Salvia gilliessi* is active against *Trypanosoma cruzi*. Acta Trop 98:118–124. doi:10.1016/j.actatropica.2005.12.007

Saxena S, Pant N, Jain DC, Bhakuni RS (2006) Antimalarial agents from plant sources. Curr Sci 85:1314–1329

Schmidt TJ, Brun R, Willuhn G, Khalid SA (2002) Anti-trypanosomal activity of helenalin and some structurally related sesquiterpene lactones. Planta Med 68:750–751. doi:10.1055/s-2002-33799

Schmidt TJ, Khalid SA, Romanha AJ, Alves TM, Biavatti MW, Brun R, Da Costa FB, de Castro SL, Ferreira VF, de Lacerda MV, Lago JH, Leon LL, Lopes NP, das Neves Amorim RC, Niehues M, Ogungbe IV, Pohlit AM, Scotti MT, Setzer WN, de N C Soeiro M, Steindel M, Tempone AG (2012a) The potential of secondary metabolites from plants as drugs or leads against protozoan neglected diseases—part I. Curr Med Chem 19:2128–2175

Schmidt TJ, Khalid SA, Romanha AJ, Alves TM, Biavatti MW, Brun R, Da Costa FB, de Castro SL, Ferreira VF, de Lacerda MV, Lago JH, Leon LL, Lopes NP, das Neves Amorim RC, Niehues M, Ogungbe IV, Pohlit AM, Scotti MT, Setzer WN, de N C Soeiro M, Steindel M, Tempone AG (2012b) The potential of secondary metabolites from plants as drugs or leads against protozoan neglected diseases—part II. Curr Med Chem 19:2176–2228

Steverding D, Tyler KM (2005) Novel antitrypanosomal agents. Expert Opin Investig Drugs 14:939–955. doi:10.1517/13543784.14.8.939

Taleb-Contini SH, Salvador MJ, Balanco JMF, Albuquerque S, de Oliveira DCR (2004) Antiprotozoal effect of crude extracts and flavonoids isolated from *Chromolaena hirsuta* (Asteraceae). Phytother Res 18:250–254. doi:10.1002/ptr.1431

Tasdemir D, Brun R, Perozzo R, Dönmez AA (2005) Evaluation of antiprotozoal and plasmodial enoyl-ACP reductase inhibition potential of Turkish medicinal plants. Phytother Res 19:162–166. doi:10.1002/ptr.1648

Tasdemir D, Kaiser M, Brun R, Yardley V, Schmidt TJ, Tosun F, Rüedi P (2006) Antitrypanosomal and antileishmanial activities of flavonoids and their analogues: in vitro, in vivo, structure-activity relationship, and quantitative structure-activity relationship studies. Antimicrob Agents Chemother 50:1352–1364. doi: 10.1128/AAC.50.4.1352-1364.2006

Taylor WR, White NJ (2004) Antimalarial drug toxicity: a review. Drug Saf 27:25–61

Tempone AG, Sartorelli P, Teixeira D, Prado FO, Calixto IARL, Lorenzi H, Melhem MSC (2008) Brazilian flora extracts as source of novel antileishmanial and antifungal compounds. Mem Inst Oswaldo Cruz 103:443–449

Thomas TRA, Kavlekar DP, LokaBharathi PA (2010) Marine drugs from sponge-microbe association—a review. Mar Drugs 8:1417–1468. doi:10.3390/md8041417

van Agtmael MA, Eggelte TA, van Boxtel SJ (1999) Artemisinin drugs in the treatment of malaria: from medicinal herb to registered medication. Trends Pharmacol Sci 20:199–205. doi:10.1016/S0165-6147(99)01302-4

Varughese G, Sabulal B, Anil J (2010) Ethnomedicinal plants in parasitic infections. In: Chattopadhyay D (ed) Ethnomedicine: a source of complementary therapeutics, pp 53–116, ISBN 978-81-308-0390-6

Vonthron-Sénécheau C, Weniger B, Ouattara M, Tra Bi F, Kamenan A, Lobstein A, Brun R, Anton R (2003) In vitro antiplasmodial activity and cytotoxicity of ethnobotanically selected Ivorian plants. J Ethnopharmacol 87:221–225. doi: 10.1016/S0378-8741(03)00144-2

Weiss CR, Moideen SVK, Simon L, Croft SL, Peter J, Houghton PJ (2000) Activity of extracts and isolated naphthoquinones from *Kigelia pinnata* against *Plasmodium falciparum*. J Nat Prod 63(9):1306–1309. doi: 10.1021/np000029g

White NJ (2004) Antimalarial drug resistance. J Clin Invest 113:1084–1092. doi:10.1172/JCI200421682

WHO (2001) Antimalarial drug combination therapy: report of a WHO technical consultation. In WHO/CDS/RBM, vol 2001.35. World Health Organization, Geneva

WHO (2002) The world health report 2002: reducing risks, promoting healthy life. World Health Organization, Geneva

Wilkinson SR, Kelly JM (2009) Trypanocidal drugs: mechanisms, resistance and new targets. Expert Rev Mol Med 11:1–31. doi: http://dx.doi.org/10.1017/S1462399409001252

Wong IL, Chan K-F, Burkett BA et al (2007) Flavonoid dimers as bivalent modulators for pentamidine and sodium stiboglucanate resistance in *Leishmania*. Antimicrob Agents Chemother 51(3):930–940. doi: 10.1128/AAC.00998-06

Woodrow CJ, Haynes RK, Krishna S (2005) Artemisinins. Postgrad Med J 81:71–78. doi:10.1136/pgmj.2004.028399

Woster PM (2009) Principles of pharmacotherapy 3: infectious diseases and disease of the respiratory tract. In Chemistry of antiparasitic agents. http://www.acsmedchem.org/module/antiparasitic.html

Wright CW (2005) Traditional antimalarials and the development of novel antimalarial drugs. J Ethnopharmacol 100:67–71. doi: 10.1016/j.jep.2005.05.012

Wright AD, Papendorf O, Konig GM (2005) Ambigol C and 2,4-dichlorobenzoic acid, natural products produced by the terrestrial cyanobacterium *Fischerella ambigua*. J Nat Prod 68:459–461. doi:10.1021/np049640w

Yabu Y, Yoshida A, Suzuki T et al (2003) The efficacy of ascufuranone in a consecutive treatment on *Trypanosoma brucei brucei* in mice. Parasitol Int 52:155–164. doi:10.1016/S1383-5769(03)00012-6

Zhang L, An R, Wang J, Sun N, Zhang S, Hu J, Kuai J (2005) Exploring novel bioactive compounds from marine microbes. Curr Opin Microbiol 8:276–281. doi:10.1016/j.mib.2005.04.008

Ziegler HL, Hansen HS, Stœrk D, Christensen SB, Hägerstrand H, Jaroszewski JW (2004) The antiparasitic compound licochalcone A is a potent echinocytogenic agent that modifies the erythrocyte membrane in the concentration range where antiplasmodial activity is observed. Antimicrob Agents Chemother 48:4067–4071. doi:10.1128/AAC.48.10.4067-4071.2004

Chapter 2
Parasitic Helminths of Humans and Animals: Health Impact and Control

Abstract Organic compounds from terrestrial and marine organisms have been used extensively in the treatment of many diseases and serve as compounds of interest both in their natural form and as templates for synthetic modifications. This chapter summarizes the present knowledge about anthelmintic effects of the extracts and some already purified natural compounds isolated from the lower marine organisms including bacteria, sponge, fungi, and algae as well as the higher plants. A brief summary on anthelmintics in use is also included to provide a background for the comparison of effective concentrations, mode of actions, and weaknesses in therapy. The main focus is placed on in vitro and in vivo activities of secondary plant metabolites (alkaloids, essential oils, flavonoids, saponins, amides, enzymes, condensed tannins, and lactones with endoperoxide bridge-artemisinins) against nematodes, trematodes, and cestodes of medical and veterinary importance, and experimental model infections. Several issues are highlighted; the synergistic effect of a number of bioactive components in plant extracts, multiple putative target sites in helminths for some of secondary plant metabolites, probably different from those of current anthelmintics, which is suggested by their modified mode of actions.

Keywords Helminths · Natural compounds · Drug discovery · Marine organisms · Terrestrial plants · Secondary plant metabolites · Anthelmintic activity

2.1 Anthelmintic Drugs: Mode of Action, Efficacy and Resistance

Anthelmintics are drugs that are used to treat infections with parasitic worms. This includes both flatworms, e.g., trematodes and cestodes and roundworms, i.e., nematodes. In the past, all the drugs used for humans were developed initially in

G. Hrckova and S. Velebny, *Pharmacological Potential of Selected Natural Compounds in the Control of Parasitic Diseases*, SpringerBriefs in Pharmaceutical Science & Drug Development, DOI: 10.1007/978-3-7091-1325-7_2, © The Author(s) 2013

response to the considerable market for veterinary anthelmintics in high- and middle-income countries (Geary et al. 2010). Discovery of benzimidazoles (BZs), a very effective broad-spectrum group of anthelmintics dates back to 1961, when thiabendazole was synthesized. The subsequent cascade of patents during the next 25 years led to the experimental or commercial development of a further 15 BZs and central to their success is their selective toxicity for helminths (Lacey 1990). The most frequently used benzimidazoles in human medicine are albendazole and mebendazole. Discovery of ivermectins as natural compounds has enlarged the group of anthelmintic drugs indicated primarily for use in veterinary medicine (for details see Sect. 2.2). All of these drugs have been discovered by means of high throughput screening of a library of synthesised chemical compounds. Once the high anthelmintic activity and low toxicity of these molecules have been demonstrated, usually the next step is the evaluation of the mechanisms of action on parasites. Various issues relating to current anthelmintic drugs, such as in vitro drug effects, in vivo efficacy, pharmacokinetic characteristics, and mode of actions have been the subjects of numerous papers. The short overview of anthelmintic drugs included in this book should serve as background for the following chapter, where concentrations, activities, and putative target sites of natural compounds are often compared with the reference drugs. There are many excellent reviews dealing with these topics, for example: Keiser and Utzinger (2010), Holden-Dye and Walker (2007), Geary et al. (2010), McKellar and Jackson (2004), Frayha et al. (1997), Martin (1997), and others.

Drug treatment necessitates a thorough knowledge of the life cycle of the parasite as well as its physiology and biochemistry and surface of the worm (tegument or cuticle). Worm surface of flatworms as well as mouth in trematodes and gut in nematodes are the key factors in drug absorption. According to the target sites in helminths which are affected by anthelmintics, they form the following groups: *Nicotinic agonists* (e.g.: levamisole, pyrantel, morantel), *Acetylcholinesterase inhibitors* (haloxon, dichlorvos), *GABA agonist* (piperazine) and *GluCl potentiators* (avermectins, moxidectin, milbemycin D), *Calcium permeability increase* (praziquantel), *β-tubulin binding* (benzimidazole carbamates), *Proton ionophores* (for example: closantel, rafoxanide, niclosamide), *Inhibition of malate metabolism* (diamphenetide), *Inhibition of phosphoglycerate kinase* (clorsulon), and *Inhibitor of arachidonic acid metabolism* (diethylcarbamazine) (Martin 1997). However, no single drug available today has use for the treatment or prevention of both nematode and trematode infections in humans (Table 2.1) and there are also differences in susceptibility to individual chemical derivates in the same class. The most important, in terms of the extent of their application in human and veterinary medicine and efficacy, are the following classes, for which we provide a concise description of their activities.

Nicotinic agonists act selectively as agonists at synaptic and extrasynaptic nicotinic acetylcholine receptors on nematode muscle cells and produce contraction and spastic paralysis. These anthelmintics have been shown to increase the membrane conductance and depolarize the membrane by opening non-selective cation ion-channels that are permeable to both Na^+ and K^+. Levamisole, pyrantel,

Table 2.1 The key drugs registered for the treatment of parasitic worms in humans (adopted from Holden-Dye and Walker 2007)

Parasitic infection	Anthelmintic drugs
Schistosomiasis (blood fluke)	Antimonials, metrifonate, oxamniquine, praziquantel
Cestodiasis (tape worm)	Niclosamide, benzimidazoles, praziquantel,
Fascioliasis (liver fluke)	Praziquntel, closantel, (and halogenated salicylamides)
Intestinal round worms	Piperazine, benzimidazoles, morantel, pyrantel, levamisole, avermectins and milbemycins, closantel (and halogenated salicylamides) emodepside
Filariasis (tissue round worms)	Diethylcarbamazine, suramin, ivermectin

morantel, and oxantel are large organic cations, and could enter the nicotinic ion-channel from the outer cell membrane. Once they pass through these channels, they produce the block at the narrow region of the channels, which subsequently cause muscle contractions in nematodes. This leads to worm paralysis in a contractile state and, once rendered immobile, the worms are expelled (Martin and Robertson 2007).

Acetylcholinesterase inhibitors are selective organophosphorus anticholinesterases. The mode of action of these compounds is to block the action of the parasite enzyme, acetylcholinesterase, leading to the excessive build-up of the neurotransmitter, acetylcholine. Metrifonate, an organophosphorus compound, is rapidly absorbed after oral administration and transformed non-enzymatically to its active metabolite, dichlorvos. The drug displays its activity exclusively on *Schistosoma hematobium,* where it inhibits cholinesterase and acetylcholinesterase to produce reversible paralysis. The paralyzed worms quickly release their hold in the bladder veins of the host and are carried eventually to the lungs where they are trapped and encased and then die. The drug is more specific for helminth cholinesterase than mammalian cholinesterases.

The group of drugs acting as *GABA antagonists and GluCl channel potentiators,* which is the mode of action of avermectins has been extensively investigated. It was shown that drugs in this group act on the same receptor as the GABA neurotransmitter in nematodes that is a ligand-gated Cl^- channel found on the synaptic and extrasynaptic membrane of nematode muscle. Ivermectin is a macrocyclic lactone derivative of avermectin-B isolated from natural source and acts as γ-aminobutyric acid (GABA) antagonist in nematodes. Avermectins are involved in the opening of the GABA-dependent chloride channels, inducing release of this transmitter which leads to the complete paralysis and immobilization of the worms (see for review: Prichard et al. 2012). The avermectins also have a receptor-mediated effect on glutamate-gated chloride (GluCl) ion channels, which has been directly correlated to nematocidal activity and which is now considered as their major mode of action (Martin and Pennington 1988). P-glycoproteins, which have a significant role in transmembrane exclusion of avermectins from the CNS, gut, and hepatobiliary tract of hosts, could account for reduced oral bioavailability of some avermectins. This effect is selective only to nematodes and arthropods, and cestodes and trematodes are not susceptible to ivermectin. Due to very low toxicity and high efficacy, ivermectins became the drugs of choice for the treatment of onchocerciasis in humans.

Benzimidazole carbamates have a broad spectrum of activities (vermicidal, ovicidal, and larvicidal activity) against many parasitic roundworms and several flatworm species, but they are not effective, for example on *Schistosoma* spp. Most human intestinal and systemic nematodes as well as systemic cestodes are susceptible to one or more of the benzimidazole compounds (Frayha et al. 1997). The most frequently used benzimidazoles in human medicine are albendazole and mebendazole (EMEA 1997, 1999). Benzimidazoles bind to intracellular tubulin, preferentially affecting parasites, thus inhibiting the formation of microtubules. This subsequently leads to disruption of cell homeostasis due to the impaired transport of secretory granules and enzymes in the cytoplasm. The mechanism of action of albendazole is by blocking glucose uptake in larval and adult stages of susceptible parasites, and also depleting their glycogen reserves, thus decreasing ATP formation (Martin 1997). The drugs are relatively insoluble in water and partially soluble in most organic solvents what has an impact on their bioavailability in tissues. Whereas efficacy is high on gastrointestinal helminths, their limited absorption and rapid metabolism means that high and/or prolonged doses are effective in the treatment of human systemic infections (Dayan 2003).

Praziquantel (PZQ) acts primarily in the tegument, where it induces a Ca^{2+} influx and rise in intra-tegumental calcium leading to the increased concentration of Ca^{2+} in sarcoplasmic reticulum of muscle cells. Disturbance in calcium homeostasis causes immediate and paralytic muscular contractions, followed by death and expulsion of parasites. The possible target of this drug, at least in schistosomes, was proposed to be voltage-gated Ca^{2+} channels, which are important regulators of calcium homeostasis in excitable cells (Kohn et al. 2003). PZQ is practically insoluble in water, and partially soluble in ethanol and organic solvents. It has remarkable range of activity against trematodes, cestodes, and is the drug of choice for all *Schistosoma* species (Watson 2009). At the standard doses, no toxicity of PZQ has been recorded and extensive studies did not show mutagenic potential for humans (EMEA 1999). It is well tolerated by patients and the major weakness of PZQ is its low efficacy against juvenile schistosomes and larval stages of cestodes (Cioli and Pica-Mattoccia 2003).

Understanding *drug resistance* is important for optimizing and monitoring, control, and reducing further selection for resistance. After decades of mass application of benzimidazole carbamates in the control of gastrointestinal nematodes in livestock, resistant strains have emerged all over the world. Recently, resistance has extended also to ivermectin and presents serious problem for the livestock industries and severely limits current parasite control strategies in humans (Prichard et al. 2012). The ATP-binding cassette (ABC) superfamily of proteins comprises several ATP-dependent efflux pumps involved in transport of toxins and xenobiotics from cells, which is essential for many cellular processes and are associated with development of multidrug resistance. P-glycoprotein (Pgp) and multidrug resistance-associated proteins (MDRP) represent two classes of these ABC transporters. They contribute to resistance to a number of

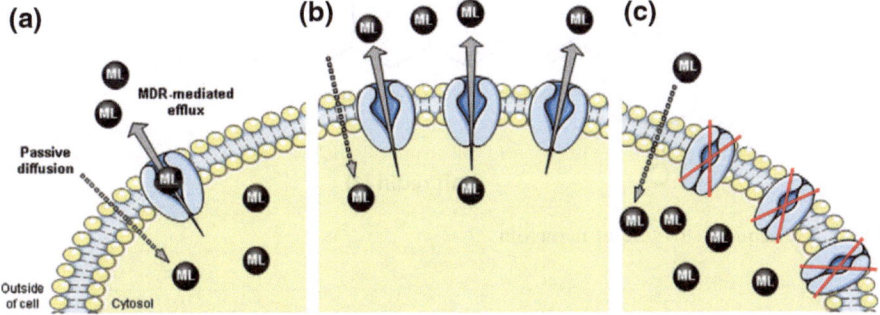

Fig. 2.1 Multidrug transporter-mediated efflux of drug, its contribution to the development of resistance and mechanism of reversion. *a* Constitutive expression of MDR transporters on cell membranes of target organism: basal efflux of drugs (example with macrocyclic lactones). *b* Overexpression of MDR transporters in response to drug pressure: increased efflux of drug and development of resistance. *c* Inhibition of MDR-mediated efflux with MDR reversal agents: enhancement of drug concentration and toxicity into the target cells after: Lespine et al. (2012)

anthelmintics, including macrocyclic lactones including ivermectin and moxidectin (see for review: Lespine et al. 2012) (Fig. 2.1). The high level of expression of MDRP to praziquantel was found in juvenile *Schistosoma mansoni* (Kasinathan et al. 2010). Several anthelmintics are inhibitors of these efflux pumps and appropriate combinations can result in higher treatment efficacy against parasites and reversal of resistance. There is the possibility that molecules with similar inhibitory action on multidrug resistance transporters will be present among a high number of naturally occurring compounds as indicated in a few studies reported in the following paragraphs.

2.2 Natural Compounds from Lower Terrestrial and Marine Organisms in Anthelmintic Drug Discovery

Marine-derived small molecules (MDSMs) from invertebrates comprise an extremely diverse and promising source of compounds from a wide variety of structural classes. They have been derived from marine plants, animals, algae, fungi, and bacteria and in total 106 marine chemicals discovered until 2002 were listed and characterized in the review of Mayer and Hamann (2005). Of these, 56 isolated marine chemicals showed one or more of anthelmintic, antibacterial, anticoagulant, antiprotozoal, antiplatelet, antituberculosis, or antiviral activities. The parasitic diseases caused by helminths and protozoan parasites that could be targets for the discovery of MDSMs were discussed in the review written by Crews and Hunter in 1993. They pointed out that very few antiparasitic drugs currently used for a spectrum of diseases were discovered after 1990 and stressed the great

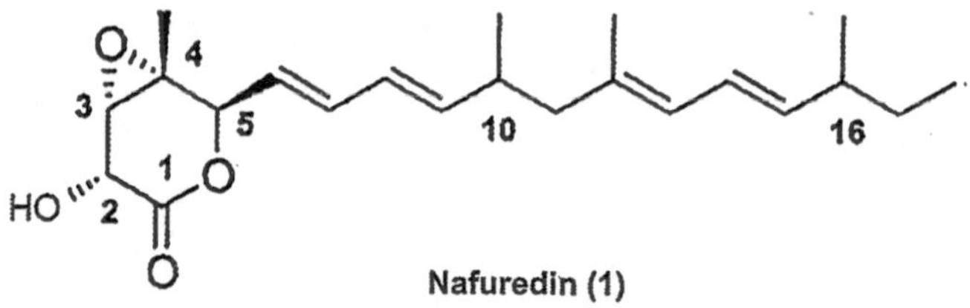

Nafuredin (1)

Fig. 2.2 Chemical structure of nafuredin

potential of MDSMs. In the recent review of Watts et al. (2010) six important parasitic diseases which affect the health and lives of over one billion people worldwide were selected and discussed in relation with natural products-based discovery. Included is a brief description of 133 marine-derived compounds displaying LC_{50}/IC_{50} below 30 μM, which were active against one or more selected parasites. The important issue highlighted in this and other reviews (e.g. Kita et al. 2007), that the majority of compounds exerted activity on protozoan species and much less invertebrate-derived molecules, were shown to be active against helminths.

Sponges (phylum:Porifera) are evolutionary ancient metazoans that have existed for 700–800 million years, populating mainly the tropical oceans in great abundance. So far about 15,000 species of sponges have been described, but their true diversity may be higher. Many species of marine sponges are associated with microbes and have been reported to produce pharmacologically active compounds (Thomas et al. 2010). An excellent example of a compound with very high anthelmintic activity which has developed from such association is *nafuredin* (Fig. 2.2).

It is chemically epoxy-δ-lactone with an olefine side chain, and was obtained from the fermentation (culture) broth of a fungal strain *Aspergillus niger* FT-0554 isolated from a marine sponge (Takano et al. 2001). Because helminths have exploited a variety of energy transducing systems in their adaptation to the peculiar habitats in their hosts, differences in energy metabolism between the host and helminths are attractive therapeutic targets for novel classes of anthelmintic compounds. NADH-fumarate reductase (NFRD) is part of a unique respiratory system in parasitic helminths (Fioravanti et al. 1998), representing a terminal electron transport system of anaerobic energy metabolism. Some kinds of parasites use this metabolism to generate ATP instead of classical glycolysis, TCA cycle, and electron transport systems. In the course of the screening of NFRD inhibitors, nafuredin was obtained and was originally tested on the nematode *Haemonchus contortus* and the tapeworm *Hymenolepis nana* in vivo (Omura et al. 2001). Sheep with *H. contortus* infection (5000 L3 larvae) were treated orally with 2 mg/kg of body weight of nafuredin. A greater than 90 % egg reduction was observed at day 11 post therapy and egg output was completely suppressed when the sheep were treated again 1 week after the first treatment. There were no signs of any side effects and no loss of body weight during the tests. The anthelmintic activity may be caused by the hampered energy metabolism of the parasite, because complex I

Fig. 2.3 NADH-fumarate reductase system of *Ascaris suum* as a target of chemotherapy. Nafuredin was found to be competitive inhibitor for rhodoquinone binding site of *A. suum* complex after Sakai et al. (2012)

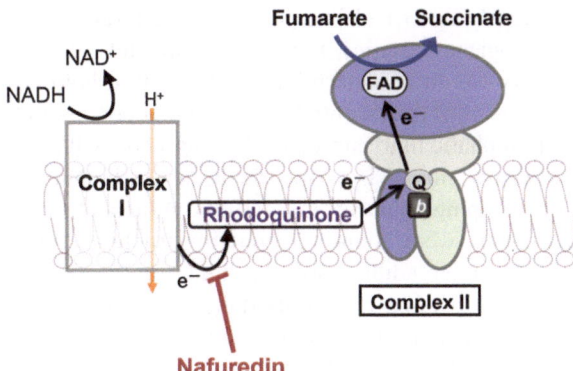

from *H. contortus* was also sensitive to nafuredin, although the inhibitions were relatively weaker than those for *Ascaris suum*.

It was shown in this study and in the study of Sakai et al. (2012) that nafuredin inhibited NFRD of nematode *A. suum* adults with an IC_{50} value of 12 nM without showing cytotoxicity for mammalian cells. Recent research on the respiratory chain of the parasitic helminth, *A. suum* has shown that the mitochondrial NADH-fumarate reductase system (fumarate respiration), which is composed of complex I (NADH–rhodoquinone reductase), rhodoquinone and complex II (rhodoquinol–fumarate reductase), plays an important role in the anaerobic energy metabolism of adult parasites. They reside in the small intestine of hosts, where oxygen tension is low. Nafuredin competes for the quinone-binding site in complex I and shows high selective toxicity to the helminth enzyme. Furthermore, nafuredin inhibits complex I (NADH-ubiquinone oxidoreductase) in L2 larvae of *A. suum*, which possess aerobic energy metabolism similar as is present in mammals at the low concentration ($IC_{50} = 8.9$ nM) (Fig. 2.3). These data demonstrated that nafuredin is effective against both adult and larval stages. In contrast, the IC_{50} value for rat liver complex I was more than 1,000 times higher than for the *A. suum* complex (Omura et al. 2001). Authors also found that harzianopyridone and the chemically related atpenins inhibit complex II (succinateubiquinone reductase -SQR) in the mitochondria of *A. suum* nematode. Complex II is indispensable for the survival of anaerobic parasitic eukaryotes and, therefore, is also regarded as a good chemotherapeutic target for novel antihelmintics. These inhibitors were isolated from *Trichoderma* sp. FTD-0795, a terrestrial fungus (Miyadera et al. 2003).

Numerous species of marine sponges have been examined for the presence of compounds with antiparasitic activity and in several species nematocidal agents were found and chemically characterized. In a southern Australian sponge of the genus *Echinodictyum* a compound (−)-*echinobetaine A* with potent anthelmintic activity was isolated and synthesized (Capon et al. 2005). The marine natural product onnamide F, isolated from the Australian marine sponge *Trachycladus laevispirulifer*, showed potent inhibition of larval development of *H. contortus* with an in vitro value of 2.6 μg/ml (Vuong et al. 2001). This compound contains a common structural motif previously described in a number of natural products exhibiting interesting phar-macological activities. For both compounds, the mechanism of their anthelmintic activity has to be determined and low threshold concentrations indicate that they interfere with the essential physiological process in helminths.

Lymphatic filariasis, caused by nematodes *Wuchereria bancrofti, Brugia malayi* and *Brugia timori,* is a parasitic disease of high medical importance and prevalence in developing countries in Southeast Asia, and sub-Saharan Africa. Several marine organisms were found to contain structures which possess antifilarial activity against either adult worms or microfilaria, or both stages. The marine *sponge Haliclona oculata* is an important source of steroids, terpenoids, alkaloids, cyclic peptides, and unsaturated fatty acids. Some of these compounds have been reported to possess diverse biological activities. Methanol extract, chloroform fraction, and one of the chromatographic fractions of this sponge revealed IC_{50} values of 5.00, 1.80, and 1.62 µg/ml, respectively, when adult *B. malayi* were exposed to these test samples for 72 h at 37 °C. Under similar exposure conditions, the IC_{50} values for microfilariae were 1.88, 1.72, and 1.19 µg/ml, respectively. The samples were found to be safe revealing >10 selectivity indices (SI) on the basis of cytotoxicity to Vero cells (monkey kidney cells). In vivo on experimentally infected gerbils, the highest efficacy (70 %) was achieved with the chromatographic fraction, where the main constituents were determined as alkaloids *mimosamycin, xestospongin-C, xestospongin-D, and araguspongin-C* (Gupta et al. 2012).

Marine alga, Botryocladia leptopoda (J. Ag.) Kyln. (order Rhodophyceae) is a red alga and this genus includes at least 24 different species which are found in the Mediterranean Sea, around South Africa, the Indian Ocean, Caribbean, around Indonesia, and in the Pacific Ocean. Antifilarial activity of the ethanol extract from this red alga was examined on a subperiodic *B. malayi* nematode, which was maintained in *Mastomys coucha*. The crude extract was active on adult worms in vitro and $LC_{100} = 125$ µg/ml was determined by a complete loss of motility. After fractionation of crude extract the activity was localized only in the n-hexane fraction and at a dose of 200 mg/kg (p.o) administered to animals for 5 days, it showed about 45 % adulticidal efficacy. Moreover, substantial proportions (71.05 %) of adult female worms were found to be sterilized (Lakshmi et al. 2004a). In other studies, very similar antifilarial effects ($LC_{100} = 125$ µg/ml) were observed on the same filarial nematode model with chloroform- methanol (1:1) extract from unidentified *green Zoanthus* (Phylum Cnidaria) and the active an-tifilarial principle was probably present in the most effective chloroform fraction obtained after fractionation (Lakshmi et al. 2004b).

With the recent emphasis of the World Health Organisation (WHO) on the development of novel antifilarial agents from natural products, screening programs involved extracts derived from both terrestrial plants and marine flora/fauna. Most of the observed microfilaricidal efficacy was slow and sustained in contrast to the dramatic microfilaricidal action of the standard drug diethylcarbamazine, indi-cating that the activity of crude extracts could be due to the combined or syner-gistic effects of more than one component. Besides, in vitro results may not always give a true picture of in vivo efficacy.

One of the most important milestones during intensive screening programs of compounds from natural sources was isolation of *avermectins* by Japanese scientists dating back to 1973 (Egerton et al. 1979). Eight active components of avermectins were isolated from the broth of lower terrestrial organism originated in a soil sample. The strain was classified as a new species of actinomycetes (bacteria) and named *Streptomyces avermectinius* (formerly *Streptomyces avermitilis*) (Takahashi et al. 2002). Avermectin activates parasite-specific glutamate-gated chloride channel (Omura Omura 2002), that leads to the neurological disruption of the parasite. Although avermectins also bind to γ-aminobutyric acid-gated (GABA) and glycine-gated chloride channels in mammals, their affinity for invertebrate receptors is more than 100 times higher.

Terrestrial *fungal species* from the genera *Trichoderma* and *Rosellinia* were identified as producers of highly specific compounds which can target an essential molecular mechanism in nematodes. For example, a fungus of the genus *Rosellinia* was identified as the source of a very effective compound (PF1022A), which is a cyclic octadepsipeptide with high nematocidal activity (Sasaki et al. 1992). *Emodepside*, a semisynthetic derivative of PF1022A was later developed by Bayer HealthCare (http://www.bayerhealthcare.com), Meiji Seika, and Astellas Pharma Inc. (http://www.astellas.com). Its target was identified as a novel 110 kDa heptahelical transmembrane receptor, named HC-110R, in nematodes, which is similar to mammalian latrophilins (Saeger et al. 2001). Latrophilins are latrotoxin specific G-protein-coupled receptors in helminths that are implicated in the regulation of exocytosis. Emodepside functions as an antagonist to latrotoxin signaling by impairing the influx of Ca^{2+}. It is highly effective against adult stages of the nematodes *Nippostrongylus brasiliensis* and *Strongyloides ratti* in rats and the nematode *Heligmosomoides polygyrus* in mice when used at an oral-dosage range of 1.0–10 mg/kg (Harder and von Samson-Himmelstjerna 2002).

2.3 Anthelmintic Potential of Higher Plants

Before commercial anthelmintics were introduced into the world market, worm infections were controlled using specific plants that, based more on belief rather than knowledge, were credited with having specific actions. Plants with antiparasitic properties could be found in temperate, tropical, as well as colder climates in the world. In traditional medicine, aqueous or powdered parts of plants were usually used at various dosages/concentrations, which could be the main reason for differences in treatment effects reported from different regions. The concentration of bioactive phytochemicals in the particular plant is also dependent on growing conditions such as climate, soil, and period of collection. Therefore the evaluation of the therapeutical potential of plant extracts must be performed in controlled in vitro and in vivo studies using established analyses and rationally designed experimental schedules.

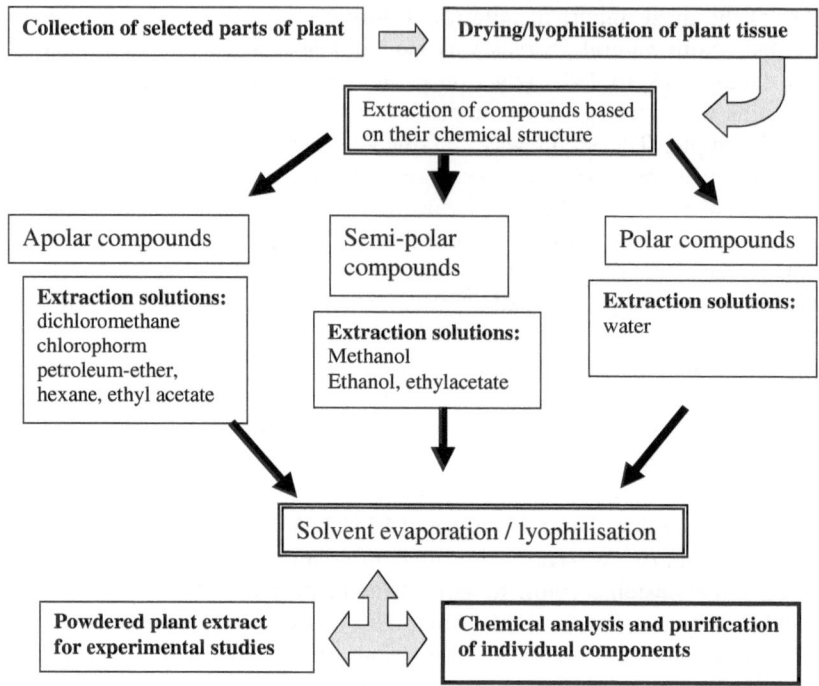

Fig. 2.4 Diagram showing the key steps of preparatory process for obtaining the single group of phytochemical/compound from the higher plants

The great advances in biochemical and analytical methods over the past 10 years allowed the separation of plant phytochemicals and consequently analysis of their chemical nature. The group of interest in this pharmacological screening program is a diverse group of secondary metabolites and usually several different classes are found together in each plant. Their basic chemical structure decides in which extraction medium (polar vs. non-polar) is dissolved in optimal quantity and quality for exerting their full bioactivity. However, many traditionally used plants also contain substances that are found to be extremely toxic. The most common extraction medium is water, however, in most studies higher therapeutical effects were observed for extracts prepared in ethanol/methanol or other organic solvents from the same plants, due to better solubility of secondary metabolites in semi-polar and apolar solutions. The schematic diagram showing the preparation of extracts for experimental testing in vitro and in vivo is shown in Fig. 2.4 and the powdered form of extract or isolated compounds is preferred. Then organic-solvent extracts are usually redissolved in DMSO, whereas water and alcoholic extracts in sterile water for further applications.

Plant-derived secondary metabolites can be divided, on the basis of their molecular formulas and structural motifs, into several classes and the most abundant are essential oils, flavonoids, alkaloids, saponins, glycosides, tannins, sesquiterpene lactones, lactones with peroxidic structure, amides, and proteins with enzymatic activity. The first step in discovering a new lead compound with desired pharmacological potential against parasites is traditionally the screening of a number of plant extracts based on their long history of medicinal applications. Countries situated in tropical regions like India, China, and other parts of Asia, South America, and Africa are especially rich in plants with a variety of bioactive molecules.

2.3.1 Plant Extracts

After an intensive search of the literature it was revealed that the majority of studies focussed on helminths living in the gastrointestinal tract of definitive hosts with fewer reports focussing on tissue-dwelling larval stages of helminths infecting humans. More than one billion people are infected with gastrointestinal helminths, and these infections are more common in the tropics where poor hygienic conditions and poverty increase the risk of infection (Brooker et al. 2006; Hotez et al. 2008). Infections of animals with gastrointestinal nematodes (GIN) are highly prevalent in both temperate and tropical areas and represent a major threat for livestock production. Nevertheless, the largest number of plant extracts has been examined for their effect to significantly reduce or remove gastrointestinal nematodes of livestock (for example see reviews: Athanasiadou and Kyriazakis 2004; Hoste et al. 2006, 2012; Hoste and Torres-Acosta, 2011; Githiori et al. 2006; de Gives et al. 2012). Extracts from a wide range of higher plants or trees have been examined so far and we summarized the data about extracts with the significant anthelmintic activity and available composition of phytochemicals. In some of these reports information about the toxicity for the hosts were available.

2.3.1.1 Plants with Activity on Cestodes and Trematodes

In the screening of compounds, which would be effective against flatworms with medical and veterinary importance, the surface structures of the tegument of cestodes and trematodes represent the potential target sites as the small molecules can be absorbed in the tegument. In experimental studies, the model infections where the life cycle can be maintained in rodents and invertebrate intermediate hosts are preferred for evaluation of higher number of plant extracts. Cestode *Hymenolepis diminuta* is the parasite of rats where adults live in the intestine and release gravid segments containing eggs. In the invertebrate hosts (beatle) cysticercoids develop from eggs and after their ingestion by rats, the immature stages and adults develop. This cestode model is widely used for many research purposes, including evaluation of cestocidal effects of plant extracts.

India is a country rich in plants, which have a long history of traditional medicinal use, including expulsion of intestinal cestodes. The extracts from leaves or plants of *Adhatoda vasica, Trifolium repens, Solanum myriacanthum*, and *Acacia* spp. were considered for a long time to contain substances with anticestodal activity. For in vivo effects, the production of eggs per gram of feces (EPG) by adult worms and a number of surviving immature and adult stages are common and suitable criteria. In the study of Yadav and Tangpu (2008) the efficacy of methanolic leaf extracts of *Adhatoda vasica* was evaluated using immature (larval) and mature stages of *H. diminuta*-albino rat experimental model. The extracts at two daily doses of 800 mg/kg reduced the EPG counts by 79.6 %, adult worm recovery was 16 %, and recovery of juvenile stages was 20 %. Effect was dose-dependent at the dose from 100 mg up to 3,200 mg/kg per os (p.o.) at which no mortality or any adverse signs with regard to body temperature or food uptake up to 72 h post therapy (p.t.) were observed. It was suggested that the anticestodal activity may be attributed to the two major constituents, the *alkaloids: vasicine* and *vasicinone*. In the work of Tangpu et al. (2004) the methanolic extract of *Trifolium repens* at a dose of 500 mg/kg reduced the mean number of excreted eggs per gram of feces (EPG) by 54.9 % and worm recovery by 40 %. In this study PZQ in the recommended dose of 5 mg/kg was only slightly more effective with an EPG reduction of 65.9 %. A reduction in EPG counts implies that phytochemicals present in extracts contributed to a higher elimination of adult worms from the intestine or to inhibition of egg production. *Solanum myriacanthum* Dunal is a perennial shrub that is used in Indian folk medicine. An oral dose of 800 mg/kg of extract, given for 3 days showed 60.49 % reduction in the EPG counts and 56.60 % reduction in the worm counts in the extract-treated group as compared to untreated controls (Yadav and Tangpu 2012). The effects of the extract were more apparent on the adult stages than on larval or immature stages of the parasite. It was assumed that the anthelmintic efficacy of plant extract may be due to the presence of secondary metabolites, particularly the *alkaloids*. Solanaceae is known for possessing a diverse range of alkaloids, like solasodine, solakhasanin, solamargine, and khasinin (Weissenberg 2001). Recently, Kamaraj and Rahuman (2011) studied the in vitro larvicidal and ovicidal activity of leaf and seed extracts of yet another *Solanum* species, i.e., *Solanum torvum* on nematode *H. contortus*. At the maximum concentration tested (50 mg/kg), a 100 % inhibition of egg hatching and larval development was recorded for an ethyl acetate extract of the plant. The extract also showed its antiparasitic effects on some hematophagous parasites of cattle and goat and also against a digenean fluke of sheep, namely *Paramphistomum cervi*. Many previous studies have assigned the antiparasitic effects of medicinal plants to these alkaloids (Athanasiadou and Kyriazakis 2004).

The anticestode activity on *H. diminuta* in the albino rat model had methanol leaf extract of *Strobilanthes discolor* (Acanthaceae). Tangpu and Yadav (2006) reported that a bioactive substance present in this plant at the dose of 800 mg/kg administered twice daily for 3 days resulted in the complete elimination of immature stages in treated rats. The doses up to 2000 mg/kg, given p.o. showed no mortality or any adverse signs in the animals, but so far the active plant component has not been determined. In the same experimental model and treatment design, the similar high activity against immature stages was obtained with methanol leaf extracts from *Zanthoxylum rhetsa* DC (Yadav and Tangpu 2009). Treatments with extracts from *Z. rhetsa* resulted in 86.6 % reduction of immature stages and EPG counts dropped to zero. This anticestodal property could be attributed to plant components *terpenoids* (xantyletin, sesamin), *alkaloids, flavonoids, and essential oil* (sabinene), which have been described as the key constituents of this plant.

Anticestode effects were demonstrated also for ethanolic extract of stem barks from *Acacia oxyphylla* on the fowl gastrointestinal cestode *Raillietina echino-bothrida* in vitro (Roy et al. 2007) and in vivo using extracts from *Acacia auriculiformis* against *H. diminuta* (Ghosh et al. 1996). Extracts decreased the motility of worms, induced distinct tissue damage in the subtegumental and somatic muscle layers, and mortality in a dose-dependent manner at concentrations between 0.5 and 20 mg/ml in vitro, indicating that condensed *tannins* as well as *saponins*, components found in herb extracts, can interact with the molecules in the tegument of cestodes. The cestode treated in vivo with 20 mg/kg showed irreversible destruction throughout the general topography of body, disorganization of the tegumental morphology, and deformation of microtriches (Table 2.4).

In recent studies, a very promising anthelmintic effect in vitro exerted extracts from plants found in tropical climates: *Lysimachia ramosa, Olea europaea,* and *Satureja khuzestanic* (Challam et al. 2010; Zibaei et al. 2012). The adult trematode, *Fasciolopsis buski*, nematode, *Ascaris suum* and a cestode, *Raillietina echinobothrida* were exposed to concentrations of 5–50 mg/ml of an alcoholic extract of *Lysimachia ramosa* Wall. Treated parasites revealed complete inactivation and loss of motility/flaccid paralysis that was followed by death at varying periods of time and deformity to the surface architecture of the worms was observed. The pharmacologically active components responsible for these effects were triterpenoids, *saponins, organic acids, and flavones*, which were recorded from different species of the genus *Lysimachia*. The protoscolicidal activity on *Echinococcus granulosus* in vitro was demonstrated for aqueous extracts of *Olea europaea* and ethanolic extracts from leaves of *Satureja khuzestanic*, which showed higher protoscolicidal activity than the aqueous extracts from *O. europaea* leaves. Loss of viability of protoscoleces was associated with a profound and characteristic morphological alteration to the surface of larvae.

Several studies showed that also extracts from plants grown to serve as human food for example: *coconut, onion, garlic, fig, date, annanas, chicory* have high anthelmintic potential against intestinal nematodes, cestodes, and trematodes. Active compounds were extracted into either chloroform, water, or polyethylene glycol/propylene carbonate (PEG/PC) and were examined on the cestode models *Hymenolepis diminuta, Hymenolepis microstoma, Taenia taeniaeformis,* and the trematode models *Fasciola hepatica and Echinostoma caproni* (Abdel-Ghaffar et al. 2011). Of all extracts tested, it was found that single extract had a very low anthelmintic effect in vivo; however, treatment of infected animals with a combination of onion and coconut extracts in PEG/PC eliminated all cestodes. The same composition of extracts was effective against *E. caproni* but failed to kill the liver fluke *F. hepatica* in the final hosts. In contrast, total or partial failure of onion oil extracts and coconut extracts given alone to kill

Table 2.4 Anthelmintic effects of plant extracts on intestinal cestode *Hymenolepis diminuta* and/or *Hymenolepis microstoma*, resp. in albino rat model

Name of plant (extraction medium)	The most effective concentration/dose of extract in selected in vitro test or in vivo studies	Main plant secondary metabolites	References
Acacia auriculiformis (ethanol)	300 mg/kg, one dose = 100 % reduction of AWN	Triterpinoid saponins: Acaciaside-A Acaciaside-B	Ghosh et al. (1996),
Onion bulbs (ethanol) + coconut powder	500 mg/kg, 8 daily doses (4 g/kg) + 500 mg/kg, 8 daily doses (4 g/kg) = 100 % reduction AWC	Allicin, Essential oils	Abdel-Ghafar et al. (2011)
Trifolium repens (methanol)	200 mg/kg, 5 doses (total 1 g/kg): FECR = 47 % 500 mg/kg, 5 doses (2.5 g/kg): FECR = 65.9 %	Not determined	Tangpu et al. (2004)
Zanthoxylum rhetsa (methanol)	100 mg/kg, 3 doses (total *0.3 g*): FECR = 23.3 %, reduction AWN = 40 % 800 mg/kg, 3 doses (2.4 g/kg): FECR = 100 %, reduction AWN = 86.6 %	Terpenoids, alkaloids, flavonoids	Yadav and Tangpu (2009)
Solanum myriacanthum (methanol)	800 mg/kg, 3 doses (total 2.4 g/kg): FECR = 60.4 %, reduction AWN = 56.6 % LD_{50} = 3.0 mg/kg in vitro	Alkaloids: solasodine, solakhasanin, solamargine, khasinin	Yadav and Tangpu (2012)
Strobilanthes discolor (methanol)	800 mg/kg, 3 doses (total 2.4 g/kg): FECR = 93.7 % reduction AWN = 90 %	Not determined	Tangpu and Yadav (2006)
Adhoda vasica (methanol)	800 mg/kg, 2 doses (total 1.6 g/kg): FECR = 79.5 % reduction AWN = 83.4 %	Alkaloids: vasicine, vasicinone, glycosides	Yadav and Tangpu (2008)

Legend
EPG eggs per gram
FECR fecal egg count reduction
AWN adult worm number

nematodes were reported by Abu-El-Ezz (2005) and Oliveira et al. (2009), indicating that anthelmintic effect of onion and coconut rely on the synergistic action of selected phytochemicals, which are able to extract with PEG/PC system and that several daily doses are necessary.

2.3.1.2 Nematocidal Activity of Plants

Parasitic nematodes represent a serious threat to humans, animals, and plants. Gastrointestinal nematode infections of livestock, which are bred for the production of meat, milk, or wool all around the world, lead to enormous economic losses. The control of these parasites has relied on the use of chemical anthelmintics, resulting in development of drug-resistant strains. Alternative control methods are biological control, vaccination, and traditional medicinal plants, which are the focus of examination over the world. The evidence of anthelmintic properties of plants is gained primarily from ethnoveterinary and ethnomedical knowledge. Novel approaches to use the plants for control of gastrointestinal nematodes in small ruminants are outlined in the excellent reviews of Githiori et al. (2006) and Hoste and Torres-Acosta (2011). In parallel with exploring non-chemical methods of control to reduce the infective larvae in the field, the search for novel pharmacologically active compounds in plant extracts is necessary for developing future anthelmintics. Obviously, in vitro assays are applied to pre-screen the activity of plant extracts and isolated components on free-living and consequently on parasitic stages of nematodes. Data collected by numerous authors to date indicate that concentrations of potentially active substances used in vitro do not always correspond to in vivo bioavailability. Therefore, in vitro assays should always be followed by in vivo controlled studies. Recently, the potential of several plant extracts with antiparasitic activities with emphasis on gastrointestinal nematodes of livestock was reviewed by de Gives et al. (2012) pointing to some other aspects. It should be taken into consideration that the different parts of higher plants (leaves, stems, roots, flowers, fruits) may contain different concentrations of the bioactive compounds that have to be determined early in the screening process. Moreover, the possible side effects on hosts should be established in vitro using cell lines or other sensitive tests to determined LC_{50} (lethal concentration at which 50 % of cells is killed), before carrying out in vivo assays.

Validation of antiparasitic activities of compounds obtained from plant extracts requires standardization of methodologies and this issue is discussed in the review of Athanasiadou et al. (2007). The authors focussed on the strengths and weaknesses of the existing methodologies used in the controlled studies and also discussed issues like the seasonal variability of the plant composition and how this can affect their antiparasitic properties. In line with their report we want to highlight the importance of identification of mechanism of action of isolated phytochemicals, which would target the unique molecular and physiological pathways of parasites.

2.3.1.3 Plants with Activity Against Gastrointestinal Nematodes

Tannin-rich plants have attracted high attention for their effect on internal nematodes in ruminants and this topic was discussed in detail in the review of Athanasiadou et al. 2001; Hoste et al. (2006, 2012); Diaz et al. (2010); Min and Hart (2003) summarized the present knowledge on antiparasitic effects of tannin-rich plants from in vivo and in vitro studies. In the majority of reports the effect was usually examined in sheep and goats, which were fed with freshly harvested forage legumes, including sulla (*Hedysarum coronariun*), sainfoin (*Onobrychis vicifolia*), trefoils (*Lotus corniculatus* and *L. pedunculatus*) and other legumes *Sericea lespedeza, Lespedeza cuneata* (for example: Bernes et al. 2000). Some studies have also tested the properties of chicory (*Chicorium intybus*) (Fig. 2.5) and are listed in

Fig. 2.5 The plants rich in condensed tannins used frequently as livestock forage

these reviews. Consumption of such bioactive plants had the beneficial effects on the host physiology and the ability to maintain homeostasis under parasitic challenge. However, the content of condensed tannins should be controlled and limited (see another part in this chapter). Such a diet has also been associated with nematodicidal effects and the most commonly reported effect was a substantial decrease in FEC (fecal egg counts), which was frequently related to significant effects on female worm fecundity. When using the whole plant extracts, these effects on nematodes can be associated with the presence of one or more plant secondary metabolites with a large range of biochemical characteristics, of which condensed tannins (CT) and sesquiterpene lactones (SL) are the most widespread in these forage legumes (Max et al. 2005). The direct role of condensed tannins on significant reduction of worm burden, hatching of eggs, and FEC in ruminants with GIN in vivo was shown after wattle tannin drenches to animals (Max 2010). Most in vitro studies reported the interference with hatching of eggs and inhibition of larval mobility for L1 and L3 stages, which can contribute to a gradual decrease in pasture contamination with infective stages. The other effect seen following incubation with tannin-rich extracts, for example, from tropical legume plant *Arachis pintoi* and *Newbouldia laevis*, was a significant inhibition of the ex-sheathment process of *H. contortus* at concentration of 1.2 and 0.6 mg/ml, respectively (von Son-de-Fernex et al. 2012; Azando et al. 2011). Feeding with chicory, which has a specific chemical composition of CT and SL resulted in the most effective "green" therapy against GIN (Tzamaloukas et al. 2005 and the above-mentioned reviews).

Several in vitro tests have been developed and proposed for the screening of drug resistant strains of nematodes (Várady et al. 2009), which target either the larval stages (larval development assay-LDA, larval migration inhibition assay-LMIA, larval escheatment inhibition assay-LEIA), the eggs (egg hatch test-EHT), or the adult stage (the adult motility inhibition assay-AMIA). Some of these bioassays are employed in screening of nematodicidal effects of plants with a record in ethnoveterinary medicine.

Haemonchus contortus is a blood-feeding abomasal nematode parasite of small ruminants and is highly pathogenic capable of causing acute disease and high mortality. The disease is characterized by hemorrhagic anemia. Resistance of this nematode to commercial anthelmintics, mainly in the tropical regions, initiated intensive research on the search of novel anthelmintic lead compounds. The effects of several plant extracts on *H. contortus* and other important gastrointestinal nematodes using EHT and LDA tests have been examined by many authors, for example: Gabino et al. (2010), Aroche et al. (2008) and others. Relevant information is summarized in Table 2.5. The anti-nematode activity was confirmed in the plant extracts from *Adhatoda vasica (L)*, *Annona squamosa*, *Cocos nucifera*, *Coriandrum sativum*, *Eucalyptus staigeriana*, *Hedera helix*, *Melia azedarach*, *Mentha piperita*, *Lippia sidoides*, *Piper tuberculatum*, *Phytolacca icosandra*, *Prosopsis laevigata*, *Spigelia anthelmia*, *Spigella torvum*. Individual extracts, isolated mostly with alcoholic and other organic solvents (n-hexane, acetone, PEG) showed the dose-dependent inhibitory effects in both tests. The highest

Table 2.5 Nematocidal effects of plant extracts on several developmental stages of *Haemonchus contortus* examined in vitro and in vivo

Name of plant	Type of study	The most effective concentration/dose of extract in selected in vitro test or in vivo studies	Plant secondary metabolites	References
Piper tuberculatum	In vitro	LD_{50} = 0.031 mg/ml in EHT LD_{50} = 0.02 mg/ml in LDA	Piplartines, piperine, essential oils,	Carvalhoa et al. (2012)
Lippia sidoides	In vitro	LD_{50} = 0.04 mg/ml in EHT LD_{50} = 0.02 mg/ml in LDA	Essential oils (76 % thymol)	
Mentha piperita	In vitro	LD_{50} = 0.037 mg/ml in EHT LD_{50} = 0.018 mg/ml in LDA	Essential oils (menthol 27 %)	
Coriandrum sativum seeds	In vitro/in vivo (sheep)	LD_{50} = 0.12 mg/ml and 100 % inhibition in EHT at 0.5 mg/ml In vivo: (0.9 g/kg 1x): significant FECR reduction and 25.5 % reduction in worm counts.	Quercetin, 3-glucuronide, linalool, camphor, geraniol, coumarins	Eguale et al. (2007a)
Eucalyptus staigeriana	In vitro	1.35 mg/ml: 99 % inhibition in EHT 5.4 mg/ml: 99 % inhibition in LDA	Essential oils	Macedo et al. (2010)
Melia azedarach	In vitro	12.5 mg/ml: 99 %inhibition in EHT 50 mg/ml: 91 % inhibition in LDA	Not known	Kamaraj et al. (2010a)
Hedera helix	In vitro/in vivo (sheep)	LD_{50} = 0.17 mg/ml in EHT In vivo: 2.2 g/kg-1x p.o. = 47.5 % reduction in FEC	Triterpenoid saponins, alkaloids, flavonoids	Eguale et al. (2007b)
Cocos nucifera	In vitro	5 mg/ml: 100 %inhibition in EHT 80 mg/ml: 99 % inhibition in LDA	Essential oils	Oliveira et al. (2009)

(continued)

Table 2.5 (continued)

Name of plant	Type of study	The most effective concentration/dose of extract in selected in vitro test or in vivo studies	Plant secondary metabolites	References
Prosopsis laevigata	In vitro/in vivo (gerbils)	20 mg/ml: 86 % larval mortality; In vivo: 40 mg/kg 1x i.p. = 42.5 % reduction of larval numbers	Not determined	Gabino et al. (2010), Aroche et al. (2008)
Annona squamosa, Solanum torvum, Terminalia chebula	In vitro	50 mg/ml: 100 % inhibition in EHT; 50 mg/ml: 100 % inhibition in LDA	Acetogenins Anthraquinones, glucopyranose, flavonoids (ellagic acid, tannic acid, gallic)	Kamaraj and Rahuman (2011)
Phytolacca icosandra	In vitro/In vivo (goats)	0.15 mg/ml: 76 % inhibition in EHT; 3 mg/ml: 67 % inhibition in LIM; LD_{50} = 0.28 mg/ml in EHT; 0.9 mg/ml: 90 % inhibition in EHT; 250 mg/kg 2x p.o. = 72 % reduction in FEC on day 11 p.t.	Flavonoids, steroids, terpenoids, Saponins, coumarins,	Hernández-Villegas et al. (2011), (2012)
Arachis pintoi	In vitro	1.2 mg/ml: 100 % inhibition of larval exsheathment in LEIA	Tannins, polyphenols	Von Son-de Fernex et al. (2012)
Gratylia argentea	In vitro	1.2 mg/ml: 66 % inhibition of larval migration in LMIA	Tannins, polyphenols	
Artemisia annua	In vitro/in vivo (sheep)	25 mg/ml: 99 % inhibition of LMIA; 3 g/kg 1x p.o. = 67 % reduction in FEC		Tariq et al. (2009), Iqbal et al. (2004)

activity at lowest concentration of extract seems to be a good criterion for selection of these plants for further in vivo studies and characterization of bioactive compounds. The low LC_{50} resp. LD_{50} (lethal concentration/dose) values in both EHT and LDA tests between 0.02 and 0.040 mg/ml were found for *Lippia sidoides, Piper tuberculatum, and Mentha piperita* extracts, which were rich in essential oils. These components were probably responsible for nematodicidal activity, mainly the abundant components of oils, for example thymol or menthol (Carvalhoa et al. 2012). The complete inhibition of EHT was observed at concentrations less than 0.5 mg/ml of crude aqueous and hyquercentindro-alcoholic extracts of the seeds of *Coriandrum sativum* (ED_{50} of the aqueous extract was 0.12 mg/ml). The main components of *C. sativum* were determined as flavonoids and essential oils (camphor, geraniol, coumarins) (Eguale et al. 2007a). However, significant decrease in FEC in artificially infected sheep was observed only for a dose of 0.9 g/kg given at week 4 post infections with L3 stage of larvae, reaching the efficacy of only 25 %. Both aqueous and hydroethanolic (50 % ethanol solution in water) extracts from *Melia azedarach* inhibited nearly completely the egg hatching and larval development at the dose of 12.5 mg/ml, indicating the presence of bioactive compounds with the ovicidal and larvicidal effects (Kamaraj et al. 2010a, b). Hydroalcoholic extracts from ripe fruits of *Hedera helix* significantly inhibited egg hatching of this nematode and ED_{50} in this test was 0.12 mg/ml and fecal egg count reduction in infected sheep treated with dose of 2.25 g/kg was 47.5 % (Eguale et al. 2007b). Similar anthelmintic activity was reported in aqueous methanolic extract from *Caesalpinia crista* on *H. contortus* (Jabbar et al. 2007). LD_{50} was achieved at the concentration of 0.134 mg/ml in EHT and 3.0 g/kg of extract given in two doses to sheep caused 93.9 % reduction in EGP. These deleterious effects of plant components were probably due to the result of blocking some important physiological processes in helminths.

Anthelmintic activity (AH) had also ethanolic extract from leaves of *Phytolacca icosandra* against *H. contortus* in EHT and LDA in vitro and in vivo in infected goats. The dose of 250 mg/kg of extract given on two consecutive days caused the highest reduction of worms (72 %) on day 11 p.t. No adverse effects were observed in all animals for the entire trial and standard methods revealed presence of saponins, coumarins, flavonoids, steroids, and terpenes as well as the presence of three fatty acids having the highest abundance (Hernandez-Villegas et al. 2011, 2012).

Many other studies have focussed on plant extracts and their activity on other gastrointestinal nematodes (GIN) of ruminants including *Teladorsagia circumcincta, Trichostrongylus axei* (all living in the abomasum), *Trichostrongylus colubriformis,* and *Nematodirus battus,* which reside in small intestine, *Trichuris ovis,* with localization in the large intestine and other GIN (*Ostertagia circumcinta, Strongyloides papillosus, Chabertia ovina*). Our search of the literature revealed that the different plants native to the country, where studies were conducted, varied in anti-nematode activity in vitro and in controlled studies in vivo. Many plant extracts given in the range of grams per kilogram of body weight of hosts failed to reach the efficacy of commercially used anthelmintic drugs, which could be attributed to the specific digestive system of ruminants.

The anti-nematode activity against GIN of sheep in vitro and/or in vivo was found in extracts of *Anacardium humile* containing mainly tannins, flavonoids, and alkaloids with $LD_{50} = 10.14$ mg/ml for the aqueous extract (Nery et al. 2010), and in the aqueous and methanolic extract from aerial parts of *Adhatoda vasica (L)*, which are rich in alkaloid vasicine, saponins, and glycosides. Different nematodes revealed different susceptibility to the same concentration of extracts in vitro and the highest effectiveness was seen at a concentration of 50 mg/ml, with inhibition between 80 and 88 % in EHT and LDA tests (Al-Shaibani et al. 2008). A significant anthelmintic effect on GIN was also demonstrated in the extracts of *Fumaria parviflora* (Papaveraceae) (Hördegen et al. 2003), in *Cichorium intybus* containing mostly tannins (Tzamaloukas et al. 2006), in *Melia azerdach* and *Trichilia claussenii*, both native to Asia, in which the main components were tannins, phenolic compounds, and steroids (Cala et al. 2012). *Agave sisalama*, sisal, is used as the source of fibers (4 % of plant) with many commercial applications. It has also a large amount of saponins composed of steroidal or triterpenoid glycosides. Silveira et al. (2012) confirmed very high effects (90–98 %) of low dose (0.12 mg/ml) of this extract in EHT and LDA tests against several GIN in vitro. A high dose of 50 mg/kg was required to reach similar anti-nematode activity in vitro with extracts from *Salvadora persica* and *Taminalia avicennoides* (Reuben et al. 2011) and chemical analysis of the aqueous extracts revealed the presence of tannins, flavonoids, saponins, sterols, terpens, and reducing sugars.

Acacia species are perennial climbing shrubs native to many Asian and African countries rich in condensed tannins. Lambs with natural GIN infection grazing *Acacia* for 4 weeks showed significantly lower mean fecal egg counts (FEC) (Akkari et al. 2008; Kahiya et al. 2003) and the anthelmintic effect of tannins against nematodes remains equivocal. The similar mode of action of tannins on nematodes as was reported for synthetic phenolic anthelmintics (oxyclozanide, niclosamide, nitroxynil) is possible, as these drugs interfere with energy generation in helminth parasites by uncoupling oxidative phosphorylation, consequently leading to depletion of parasite ATP (Martin 1997). Another report suggested tannin's ability to bind to glycoproteins on the cuticle of parasites (Hoste et al. 2006).

Artemisia species are the source of artemisinin, which was recognized as a new lead compound with high anti-protozoan effect. It was partially effective also against *Schistosoma* spp. and *Fasciola* spp. (Keiser and Utzinger 2007, see the paragraph on Sect. 2.3.8). Several studies examined the anthelmintic potential of the extracts from *Artemisia* spp. on nematodes with various levels of success. Significant inhibition (nearly 100 %) of *H. contortus* larval motility in vitro was found at the concentration of 25 mg/ml after several hours of exposure and worm paralysis and/or mortality was seen at 6 h post exposure. In sheep with this infection, and treated with a single dose of up to 3 g/kg of extract, the reduction of infection was only 67 % (Tariq et al. 2009; Iqbal et al. 2004).

In the study of Mehlhorn et al. (2011a), especially prepared extracts from *coconut* dried endosperm and onion bulbs were examined in vivo in sheep infected with various gastrointestinal nematodes and/or *Moniezia* expansa (Cestoda). A combination of onion and coconut extracts each containing 60 g of dried mass was given to sheep for 8 days combined with PEG/PC. In all cases, the worms disappeared from the feces indicating the 100 % de-worming effect. PEG alone probably improved the absorption of individual substances in garlic and coconut powders and when given alone, it had no effect on worm burden. Administration of onion powder alone showed low efficacy against worms (Klimpel et al. 2011; Abdel-Ghaffar et al. 2011). *Allium sativum (garlic)* extract exerted anthelmintic activity against *H. contortus* in vitro (Iqbal et al. 2001), but pure allicin (the leading substance in onion and garlic) remained ineffective against the worms. These studies demonstrated the synergistic effect of garlic and coconut given as powder to animals and was reviewed by Mehlhorn et al. (2011b).

High-throughput screening of chemical substances for their anthelmintic activity requires a well-established model. The use of parasitic nematodes as a screening system in vitro is hindered by the difficulty of maintaining them in vitro for prolonged periods of time and the need to have infected animals as sources of eggs or adults. In this respect, the free-living soil nematode *Caenorhabditis elegans* was proposed as a system to test products for potential anthelmintic effect against small ruminant gastrointestinal nematodes, including *H. contortus* (Katiki et al. 2011a, b). *Leucosidea sericea* is a plant native to southern Africa and was used to expel parasitic intestinal worms (vermifuge) and as an astringent in combination with other plants. The study of Aremu et al. (2010) aimed to examine whether extracts from this plant could exert both anthelmintic and anti-inflammatory activity. Cyclooxygenase enzymes (COX-1 and -2) were used to determine the anti-inflammatory potential of the plant extracts as their production is increased in acute and chronic inflammatory conditions associated with many diseases. An in vitro colorimetric assay for the determination of free living nematode larvae viability enabled the recording of the minimum lethal concentration (MLC) values of the extracts against *C. elegans* var. Bristol (N2). At the highest screening concentration (250 µg/ml), the PE (petroleum extract) of the leaves exhibited the highest COX-1 and COX-2 inhibition of 97 and 91 %, respectively. Alkaloids and saponins were only detected in the leaf and stem extracts, respectively, of *L. sericea*. MLC values for all the organic solvent extracts of the leaves between 0.26 and 0.56 mg/ml corresponded with high anthelmintic activity.

2.3.1.4 Plant extracts with activity on nematode species of hosts with monogastric digestive systems including Heligmosomoides bakeri, Trichinella spiralis, Brugia malayi, Toxocara cati, A. suum.

Trichinellosis is a worldwide zoonotic infection caused by nematodes of the genus *Trichinella* and in nature is widespread in wild carnivorous animals (Murrell and Pozio 2011). Pigs and also humans can be infected after the ingestion of meat containing live larvae and therapy for patients is not always successful (Pozio et al. 2001) with anthelmintics, due to low bioavailability for larvae localized in the muscles. Methanol extracts from *Artemisia absinthium* and *A. vulgaris* given to rats at the dose of 300 mg/kg during enteral (adult) phase reached the rate of larval reduction in various muscles between 37 and 75 %. During the parenteral (encapsulated larvae) phase, the higher dose of 600 mg/kg decreased the larval rate between 43 and 66 % (Caner et al. 2008). These data suggest that concentration of artemisinin in extracts was not sufficient for complete worm elimination and longer administration could be the choice. The synergistic effect of other plant components like flavonoids is also possible.

Heligmosomoides bakeri infection in rodents is a suitable experimental model for human gastrointestinal nematodes. Extracts from leaves of *Ageratum conyzoides* (Asteraceae), used in ethnomedicine in Africa, showed 53.3 % inhibition of the embryonation process in eggs at concentration of 3.7 mg/ml and LC_{50} was 1.5 mg/ml for L2 stage larvae (Poné et al. 2011). Lymphatic filariasis is a major health problem in the local population in tropical and subtropical countries and the disease is caused by the filarial nematodes *Wuchereria bancrofti, Brugia malayi,* and *Onchocerca volvulus.* Treatment relies on two drugs, diethylcarbamazine and ivermectin, but they have poor activity on adult parasites. Several plant extracts have been examined for antifilarial activity in vitro and in laboratory animals. The crude extract of *Caesalpinia bonducella* showed very high effects against all stages of nematode *B. malayi* (Gaur et al. 2008) extract from stem portion of plant *Lantana camara* administered to rats at the dose of 1 g/kg for 5 days killed 43 % of adults and sterilized 76 % of the surviving female worms. Two isolated compounds, *oleanonic acid* and *oleanolic acids* might be responsible for this effect (Misra et al. 2007). All *B. malayi* worms were killed in vitro with extract at concentration of 0.031 mg/ml. Antifilarial activity on microfilaria of *B. malayi* in vitro was identified in methanolic extracts of roots of *Vitex negundo* L. and leaves of *Aegle marmelos Corr* by Sahare et al. (2008). Chromatographic analysis revealed the presence of *alkaloids, saponin,* and *flavonoids* in *Vitex negundo* and *coumarins* as the main component in plant *A. marmelos,* and both extracts at concentration of 0.1 µg/ml showed complete loss of motility of microfilaria after 48 h, indicating the inhibition of the essential physiological process in larvae. Transcuticular/tegumental diffusion is a common means of entry for non-nutrient and non-electrolyte substances into helminth parasites. It was shown that this route is predominant for the uptake of major groups of anthelmintics like benzimidazole, levamisole, and ivermectin, and also some of bioactive phytochemicals can enter tegument/cuticle in this way. Lipophilic anthelmintics have a greater capability to cross the external surface of helminths than hydrophilic compounds (Geary et al. 1999).

2.3.2 Alkaloids

Alkaloids represent a highly diverse group of compounds that are related only by the occurrence of a nitrogen atom in a heterocyclic ring. The nitrogen in the alkaloid molecule is derived from amino acid metabolism. Since the amino acid skeleton is often largely retained in the alkaloid structure, alkaloids originating from the same amino acid show similar structural features and can be classified according to their biosynthetic origin. Plants are estimated to produce approximately 12,000 different alkaloids, which can be organized into groups according to their carbon skeletal structures (Ziegler and Facchini 2008).

Alkaloids often have pronounced bioactivities and are therefore thought to play an important role in the interaction of plants with their environment. Alkaloids and extracts of alkaloid-containing plants have been used throughout human history as remedies, poisons, and psychoactive drugs. Several alkaloids are used medicinally or provide lead structures for novel synthetic drugs (Fester 2010).

A great number of alkaloids originating from marine organisms showed a potent activity on parasitic protozoa and on several helminth species in vitro and some specifically inhibit essential enzymatic systems in parasites (see for review Watts et al. 2010 and Chap 1 in this book). However there are few studies, in which alkaloids isolated from higher plants, were examined as anthelmintic agents.

Many known anthelmintics are effective against nematodes living in gastrointestinal system, but limited efficacy has been reported against larvae which migrate to human tissues where they may cause a severe pathology. This is the case for *Toxocara canis* and *Toxocara cati* larvae, both zoonotic nematode infections. Migration of larvae is associated with many non-specific pathological syndromes, eosinophilia, and an allergic type of immune response. The seroprevalence of toxocariasis in the healthy human population was estimated to be 5–10 % and the disease ranks among the most frequent parasitic diseases in temperate climates. The search for plant phytochemicals effective against second stage larvae of *T.canis* were the focus of studies of several research groups. Satou et al. (2002a) developed a specific test for the evaluation of nematocidal activity in vitro and introduced the concept of Relative Mobility (RM) of larvae. This shows the concentration of a compound at which RM equals 50 % of the control, the RM_{50}, value. Using this test, a set of *isoquinoline alkaloids* isolated from the plants *Macleaya cordata, Chelidonium majus,* and *Corydalis turtschaninovani*, grown in Japan for various purposes, were examined for inhibition of mobility of larvae and their cytotoxicity. Some of the isolated alkaloids were highly cytotoxic in HL60 tissue-culture cells, so only those with a high RM_{50}/IC_{50}, ratio (Selectivity index: SI > 100) were proposed as potential anthelmintic alkaloid molecules, namely *allocryptopine, dehydrocorydaline, and papaverine.* However, the mechanism of inhibition of larval motility is so far unclear.

In an effort to identify a treatment for *Sstrongyloides stercoralis* larval migrans in humans, the effects of *isoquinoline alkaloids* were examined by Satou et al. (2002b) using infective third-stage larvae of *Strongyloides ratti* and *S. venezuelensis* as model nematodes for *S. stercoralis.* The nematocidal activity of a set of isoquinoline alkaloids isolated from the same plants as in the previous study was further evaluated in vitro on L3 stage larvae. Total inhibition of mobility corresponding to the complete paralysis of larvae was test of choice for

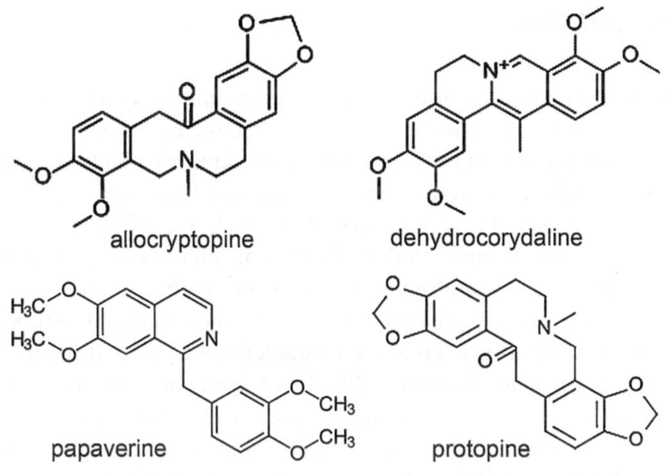

Fig. 2.6 Chemical structure of selected isoquinoline alkaloids with anthelmintic activity

comparison of nematocidal activity of all alkaloids. Three alkaloids: *protopine, d-corydaline, and l-stylopine* (Fig. 2.6) exhibited strong nematocidal activity at low concentrations (52/33 μM, 18/30 μM, 14/13μM for *S. ratti*, and *S. venezuelensis*, respectively.), and showed little cytotoxicity (SI > 100) for HL60 tissue-culture cells. However, these concentrations were significantly higher than that found for ivermectin (2.2 and 2.3 μM, resp.), which was the most effective strongyloidosis treatment.

β-carboline alkaloids are a large group of natural and synthetic indole alkaloids with different degrees of aromaticity, some of which are widely distributed in nature, including various plants, foodstuffs, marine organisms, insects, mammals as well as human tissues, and body fluids. They possess diverse biological activities and can inhibit several enzymatic systems in mammals (Cao et al. 2007).

A set of 17 different *β-carboline alkaloids* from the plants *Picrasma quassioides* and *Ailanthus altissima* which grow in Japan was examined using *T. canis* larvae in in vitro screening (Satou et al. 2005). Inhibition of mobility (larval paralysis) and low cytotoxicity were found only for a few alkaloids and only one alkaloid with no cytotoxicity at 0.1 mg/ml was examined in vivo on infected mice. In attempt to slow-down metabolism of this alkaloid, it was entrapped in pegylated liposomal carriers and this drug formulation showed higher reduction of larval numbers in the brain of mice than reference drug albendazole, as well as decreased eosinophilia. No side effects in mice were observed. Alkaloids with reported high nematocidal potential on larvae migrating in the tissues of hosts which also showed very low cytotoxicity seem to be interesting molecules. Therefore, it is worth to elucidate their mode of action on larvae and inhibition of motility suggests that their target could be in neuromuscular regulation.

2.3.3 Essential Oils

Essential oils are volatile, natural, complex compounds characterized by a strong odor and are formed by aromatic plants as secondary metabolites. In nature, essential oils play an important role in the protection of plants as antibacterials, antivirals, antifungals, insecticides, and also against herbivores by reducing their appetite for such plants. There are several methods for extracting essential oils. These may include use of liquid carbon dioxide or microwaves, and mainly low or high pressures distillation employing boiling water or hot steam.

Essential oils are very complex mixtures which can contain 20–60 components at quite different concentrations. They are characterized by two–three major components at fairly high concentrations (20–70 %) compared to others components present in trace amounts. The major component is composed of *terpenes* (for example geraniol, carvacrol, thymol, cymene, sabinene, alpha-pinene, betapinene, citronellol, sesquiterpene farnesol) or *terpenoids* (menthol, ascaridole), and the other components of *aromatic and aliphatic constituents*, all characterized by low molecular weight (for example cinnamyl alcohol, eugenol, safrole) (Fig. 2.7). In general, the abundant compounds determine the biological properties of the particular essential oil and it is likely that their mode of action involves several targets in the cells. (see for review: Bakkali et al. 2008). The hydrophobicity of essential oils enable them to pass through the cell wall and cytoplasmic membrane, disrupt the structure of their different layers of polysaccharides, fatty acids and phospholipids, rendering them permeable (Turina et al. 2006). This extensive change in fluidity/ permeability of membranes, results in the leakage of radicals, cytochrome C, calcium ions, and proteins, as in the case of oxidative stress and bioenergetic failure. Cytotoxicity to eukaryotic cells appears to include such membranes damage. In general, the cytotoxic activity of essential oils is mostly due to the presence of phenols, aldehydes, and alcohols (Sacchetti et al. 2005). Clearly, it has been shown by Bakkali et al. (2005) that the tested essential oils presented a specificity in the amplitude, but not in the mode of action, and of the biological effects, i.e., cytotoxicity, cytoplasmic mutant induction, gene induction, and antigenotoxic effects.

Essential oils or some of their constituents are effective against a large variety of organisms including bacteria, fungi, viruses, protozoa as well as metazoan parasites (Bakkali et al. 2008). Plants known for a high content of essential oils are

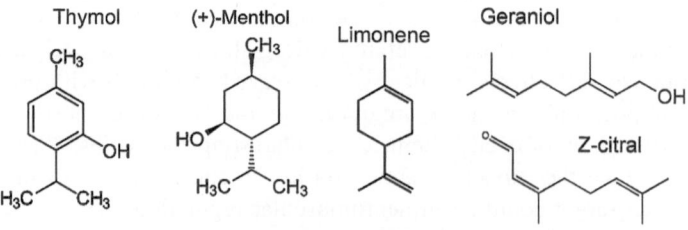

Fig. 2.7 Chemical structure of selected components of essential oils with anthelmintic activity

for example: *Origanum compactum, Coriandrum sativum, Eucalyptus staigeriana, Artemisia herba-alba, Cinnamomum camphora*, leaf and seeds of *Anethum graveolens, Mentha piperita, Salvia* spp., *Nigella sativa*, and others.

We showed in previous part of this chapter, that the extracts from several of these plants exerted significant anthelmintic activity and their essential oils probably contributed to the final effects in vitro and in vivo. To be able to study a direct effect of essential oils on helminths in vitro, their limited solubility in water requires using suitable solvents, usually non-ionic detergents (Tween 20, Tritox-X-100) or DMSO. However, the final effect on infections in animals is influenced by low absorption, the first pass effect in the liver and metabolization into individual components (Fandohan et al. 2008), each component having different mode of action. As the mode of absorption in monogastric and polygastric animals is different, this may influence also the fate of essential oils.

So far, only several essential oils have been examined for their activity on helminths in vitro and in controlled in vivo studies. Whereas some essential oils exerted nematocidal activity only in vitro (for example oil from *Cymbopogon* spp. *and Chenopodium ambrosioides*), others (for example oil from *Eucalyptus staigeriana, Ocimum gratissimum*) significantly reduced infection with GIN in ruminants and also in rodent models in vivo. Oil composition was different in these plants, which probably influenced the in vivo activity.

Herbs of *Cymbopogon* spp. belong to family Poaceae and the essential oil is used in Brasil for its characteristic aroma as an insecticide. In vitro activity of oils from *Cymbopogon schoenanthus, Cymbopogon martini*, and *Mentha piperita* was studied against developmental stages of trichostrongylidae from sheep naturally infected (95 % *Haemonchus contortus* and 5 % *Trichostrongylus* spp.) (Katiki et al. 2011a, b). The essential oil of *Cymbopogon schoenanthus* had LC_{50} value of 0.045 mg/ml in EHT, 0.063 mg/ml in LDA, 0.009 mg/ml in LFIA, and 24.66 mg/ml in larval exsheathment assay (LEA), but oil from *M. piperita* was much less effective. The major constituent of the essential oil from *M. piperita* was *menthol* (42.5 %), while for *C. martinii* and *C. schoenanthus* the main component was *geraniol* (81.4 and 62.5 %, respectively). However, in young lambs experimentally infected with a multidrug-resistant *Haemonchus contortus* strain, administration of *C. schoenanthus* essential oil (180 mg/kg and 360 mg/kg BW) for 3 consecutive days failed as an anthelmintic treatment. No statistically significant reduction in fecal egg count, packed cell volume or total worm count was observed (Katiki et al. 2012) and toxicity symptoms were observed at 360 mg/kg, with signs of discomfort, apathy, lethargy, and drowsiness in treated animals. Similarly, oil-containing extracts from *Chenopodium ambrosioides* were effective on nematodes in vitro, but short-term treatment with up to 0.4 ml/kg BW of oil was not effective in reducing the number of nematode adults or eggs in infected goats (Ketzis et al. 2002). The major oil component was detected as the terpenoid *ascaridol*, which in the same way as the terpenoid *geraniol*, became probably ineffective following absorption in polygastric animals.

Eucalyptus staigeriana is most commonly used in Brazil for extraction of essential oil, in which several secondary metabolites were detected and the highest concentration was found for (+)-*limonene* (28.8 %), α-*terpinolen* (9.4 %), *Z-citral* (10.77 %), and *E-citral* (14.16 %) (Macedo et al. 2010). Using the in vitro model of *H. contortus,* oil inhibited the in vitro egg hatching test (EHT) and larval development assay (LDA) by 99.2 % at low concentration of 1.35 mg/ml and 5.5 mg/ml, respectively. In a subacute toxicity study on mice, the LD_{50} was 4 112.94 mg/kg after oral administration. Fecal egg count reduction (FECR) in goats after treatment with oils varied from 61.4 to 76.57 % at 8 and 15 days p.t. and biochemical tests indicated that kidney and hepatic functions were preserved, and this essential oil did not produce toxicity during the treatment period in experimental animals. Oil from E.citriodora at concentration of 5.3 mg/ml inhibited egg hatching of GIN by 98.8% (Macedo et al. 2011). The use of these essential oils would be justified even with effectiveness less than 95 %, especially in situations where the synthetic anthelmintic was not recommended, such as organic farms, in milk-producing animals.

Nematodicidal activity of essential oils extracted from other plants reached values between 94.5 and 100 % and various concentrations of essential oils were used in the same tests and nematode model. Essential oils were isolated from *Eucalyptus globulus* (Macedo et al. 2009), *Ocimum gratissimum* Linn. (Labideae) with the main component *eugenol* (43.7 %) and 1,8-cineole (32.71 %) (Pessoa et al. 2002), *Croton zehntneri* and *Lippia sidoides* with the main components *anethole* and *thymol* (Camura-Vasconcelos et al. 2007). The similar inhibition of egg hatching as obtained with 1.0 % thiabendazole, were observed after exposure to oil of *O. gratissimum* (0.5 %) and eugenol (1.0 %) in vitro. At concentration of 800 mg/kg, the essential oils from *C. zehntneri* and *L. sidoides* were 46.3 % and 11.64 % effective against sheep gastrointestinal nematode *H. contortus*, indicating that *thymol* and *anethole* are the probable active substances having different mechanisms of action on nematode cells. Albuquerque et al. (1995) reported that *C. zehntneri* essential oil and anethole blocked muscle contractions and reduced the response in muscle to acetylcholine implying action sites in muscle fibers. In contrast, the chemical structure of thymol implies its possible amphipathic and/or hydrophobic behavior. This suggests an ability of thymol partition in the membrane from an aqueous part as well as a capacity to affect the membrane organization and the surface electrostatics. This assumption may explain the activity of thymol on the permeability of membranes and on the activity of membrane intrinsic proteins such as ATPases or membrane receptors (Sánchez et al. 2004).

Based on present data, the promising substances in terms of therapy of nematode infections in monogastric animals seem to be oil from *Cymbopogon* spp. and *Croton zehntneri*. Significant in vitro and in vivo nematocidal activity was found also in essential oils from *Eucalyptus* spp., in small ruminants and components of these oils deserve further investigations, namely citrals, limonene, anethole, and eugenol. Desired effects can result in reduced reinfection and reduced worm loads leading to decreased pasture contamination levels (Max 2010).

The protoscolicidal effect of *thymol* was tested in vitro against *E. granulosus tapeworm* at a concentrations of 10, 5 and 1 μg/ml in medium (Elissondo et al.

2008). At a concentration of 10 μg/ml thymol reduced viability to 53.5 % after 12 days of incubation and to 11.5 % after 42 days of culture. The primary site of damage was the tegument of protoscoleces and morphological changes included contractions of the soma regions, formation of blebs on the tegument, rostellar disorganization, loss of hooks, and destruction of microtriches, which are directly associated with nutrient absorption. Blebs and alteration of mitrotriches probably interfere with protoscoleces' nutrition explaining the later appearance and gradual elevation of the toxic effect of thymol. Stimulation of motility of protoscoleces was not observed. Also in another study (Moazeni et al. 2012) thymol as the main constituent in the essential oil from the fruits of plant *Trachyspermum ammi* (50 %) was probably the active molecule responsible for significant scolicidal effect in vitro. Addition of 5 mg/ml of essential oil killed 51.89, 72.20, 88.64, and 100 % of protoscolices after 10, 20, 30, and 60 min, respectively.

The tegument of cestodes and trematodes is morphologically and physiologically different than cuticule of nematodes and due to the absorption function for nutrients, essential oils will likely to cross individual layers of tegument. Based on our literature search it seems that *Schistosoma mansoni* has been the most intensively studied *trematode* in relation to the therapeutical potential of essential oils. Although praziquantel is effective against all medically important species of genus *Schistosoma*, it is ineffective against schistosomula, which motivates the search for new active compounds. Anti-schistosomal activity against various stages of this trematode was reported for several essential oils isolated from a variety of plants, mostly in recent years (Magalhaes et al. 2012; Ael-Banhawey et al. 2007; Parreira et al. 2010; de Melo et al. 2011; Caixeta et al. 2011; de Oliviera et al. 2012; Mahmoud et al. 2002). Available composition and the concentrations of those with the highest anthelmintic activity of essential oils so far examined are summarized in Table 2.6. The important observation reported by all these studies was that the concentration of about 100 mg/ml of oil was required to achieve effects similar to PZQ at a concentration of 12.5 μg/ml in vitro. At such high concentration of essential oils from *Piper cubeba, Baccharis dracunculifolia, Baccharis trimera Ageratum conyzoides, Bidens sulphurea, and Plectranthus neochilus, Curcuma longa* oils caused the death of all adult worms and promoted separation of the couple pairs into individual male and female within 24–30 h. In vivo all of these oils given at higher doses were responsible also for remarkable reduction in the number of eggs and most of them reduced the viability of cercariae and schistosomula, an effect not seen with PZQ. Tegumental damage, destruction of tubercles and spines, and suckers of adult worms were observed in dead worms obtained from treated animals (de Oliviera et al. 2012).

Seeds of *Nigella sativa* have been employed for thousands of years as a spice. The immunomodulatory, therapeutic and anti-oxidant properties of oil and its constituents isolated from the seeds of this plant, in particular *thymoquinine* (TQ), have been reviewed by Salem (2005). Low doses (250 μl/kg) of the oil from *N. sativa* exerted trematocidal effect in vivo only when sidr honey was coadministered daily for 7 weeks post infection, whereas single therapy with oil was

Table 2.6 In vitro and in vivo anthelmintic effects of selected essential oils with determined components on nematodes (*H. contortus*), trematodes (*S. mansoni, F. gigantiga*) and cestodes (*E. multilocularis*)

Name of plant	Main constituents of essential oil	Parasitic infection	The most effective concentration/dose of oil in selected in vitro test or in vivo studies	References
Croton zehntneri	Anethole-64 % estragole < 15 %	*Haemonchus contortus*	1.25 mg/ml: 98 % inhibition in EHT 10 mg/ml:99.2 % inhibition in LDA	Camura-Vasconcelos et al. (2007)
Eucalyptus globulus	(+) limonene	*Haemonchus contortus*	21.75 mg/ml: 99.3 % inhibition in EHT 43.5 mg/ml: 98.7 % inhibition in LDA	Macedo et al. (2009)
Eucalyptus citriodora	(+) limonene	*Haemonchus contortus*	5.3 mg/ml: 98.8 % inhibition in EHT 10.6 mg/ml: 99.7 % inhibition in LDA	Macedo et al. (2011)
Eucalyptus staigeriana	(+) limonene - 28.8 % Z-citral-10.77 % E-citral-14.16 %	*Haemonchus contortus*	1.35 mg/ml: 99.2 % inhibition in EHT 5.5 mg/ml: 99.0 % inhibition in LDA	Macedo et al. (2010)
Cymbopogon schoenanthus	Geraniol-62.5 %	*Haemonchus contortus*	LC_{50} = 0.045 mg/ml in EHT LC_{50} = 0.063 mg/ml in LDA	Katiki et al. (2011a, b)
Lippia sidoides	Thymol < 50 %	*Haemonchus contortus*	20 mg/ml: 94.5 % inhibition in LDA	Camura-v Vasconcelos et al. (2007)
Piper cubeba L.	Sabinene-19 %, eucalyptol-11 % 4-terpineol-6.3 % β-pinene-5.8 %	*Schistosoma mansoni*	0.012–0.05 mg/ml: reduction of cercarie viability; no effect on adult worms; separation of coupled adut worms	Magalhaes et al. (2012)
Baccharis trimera L.	Nerolidol-33 % Spathulenol-16 %	*Schistosoma mansoni*	0.130 mg/ml: 100 % mortality of adults, peeling of tegumental surface (tubercles, spines)	de Oliviera et al. 2012
Plectranthus neochilus	β-caryophyllene - 28.2 % α-thujene-12.2 % α-pinene-12.6 %	*Schistosoma mansoni*	0.1 mg/ml for 24 h: 100 % mortality of adult worms, dose-dependent reduction in % of developed eggs	Caixeta et al. (2011)

(continued)

Table 2.6 (continued)

Name of plant	Main constituents of essential oil	Parasitic infection	The most effective concentration/dose of oil in selected in vitro test or in vivo studies	References
Nigella sativa	Diterpene alkaloids thymol, thymoquinone dithymquinone	*Schistosoma mansoni*	5.0 ml/kg daily (14 days) -32 % reduction of worm burden and reduction in eggs load in the intestine of treated mice	Mahmoud et al. (2002)
Ageratum conyzoides L.	Precocene I-74 % E-caryophyllene-14 %	*Schistosoma mansoni*	0.1 mg/ml for 120 h: 100 % mortality of adult worms, 75 % of separation of coupled adut worms	de Melo et al. (2011)
Synthetic	Thymol	*Echinococcus multilocularis*	0.010 mg/ml: 53.5 % reduction of protoscoleces viability after 12 days, massive tegumental damage	Elissondo et al. (2008)
Allium sativum (garlic)	allicin S-Allyl-L-cysteine g-glutamyl-Sallylcysteine	*Fasciola gigantica*	3 mg/ml: complete paralysis of worms after 15 min of incubation	Singh et al. (2009)
Piper longum	β-caryophyllene-33.44 % 3-carene-7.58 % Eugenol-7.39 %	*Fasciola gigantica*	3 mg/ml: initial excitation following paralysis of worms after 15 min of incubation	Singh et al. (2009)

ineffective (Mostafa and Soliman 2010). In another study on the same kind of infection, the higher daily doses (2.5 ml/kg or 5.0 ml/kg) of *N. sativa* oil administered orally to mice for 2 weeks reduced the number of *S. mansoni* worms in the liver only by 22 and 32 %, respectively, and decreased the total number of ova deposited in both the liver and the intestine (Mahmoud et al. 2002). Thymol and thymoquinine, as the main components of oil, could be responsible for observed trematocidal effect. Reductions in parasite burden and granuloma size coincided with partial amelioration of the *Schistosoma*-induced liver fibrosis and changes in ALT, GSH, AP activities in serum, suggesting that the schistosomicidal effect of *N. sativa* oil might be induced partly by its anti-oxidant effect documented in numerous studies (Ramadan et al. 2003). Similarly, treatment with *N. sativa* oil decreased the hepatocellular necrosis, degeneration, and advanced fibrosis in CCl4-induced liver fibrosis in rabbits (Türkdoğan et al. 2001). The effect of the oil could, at least partly, be attributed also to drug-induced modulation of the immune response to *Schistosome* eggs trapped in the liver as in vitro studies showed that oil enhanced the production of IL-3 by human lymphocytes and had a stimulatory effect on macrophages (Haq et al. 1995).

The relatively high effective concentration of essential oils and similar effects on flatworms indicates that individual components of oils probably did not selectively act on a parasite-specific molecular targets and that the final deleterious effect might occur due to synergism of all or some of the components.

More light onto the mechanism of action of essential oils on viability and motility of trematodes was brought by the study of Singh et al. (2009). In vitro exposure to essential oils from *Allium sativum* (garlic) and *Piper longum* (Indian long pepper) have markedly changed muscular activity of the whole worms and muscle strips of the liver fluke *Fasciola gigantica*. Essential oil from *A. sativum* caused complete paralysis of the fluke after 15 min of administration of 3 mg/ml and flaccid paralysis in the strip preparations. In contrast, essential oil from *P. longum* first induced marked excitatory effect and then the flaccid paralysis of the whole worm following 15 min exposure to the same concentration. These effects were irreversible and the rapid responses to oils suggest the involvement of neuromuscular system of worms. Many of the anthelmintics cause paralysis of helminth parasites by disrupting one or the other aspect of their neuromuscular system, but at much lower concentrations (Loukas and Hotez 2005). With regard to the effect of essential oil of *A. sativum* and *P. longum*, they produced grossly similar effect on both preparations, although the main components of both oils are different (Liu et al. 2007; Itakura et al. 2001). It was concluded that tegument did not interfere with the action of essential oils on smooth muscle actin of *F. gigantica*. Nevertheless, observations on strip preparations do not support the general assumption that the tegument provides a barrier in the translocation of drugs to neuromuscular targets in trematodes (Sobhona et al. 2000).

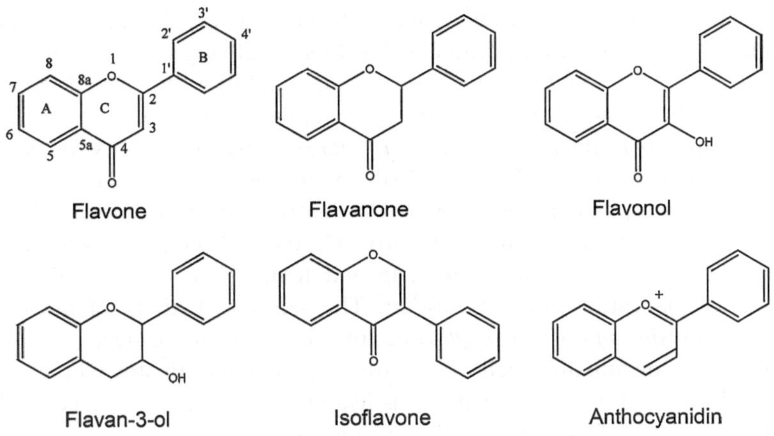

Fig. 2.8 Chemical structure of individual types of flavonoids

2.3.4 Flavonoids and Polyphenols

Flavonoids are plant pigments present in almost all terrestrial plants, where they provide UV protection and color. They are synthesized from phenylalanine and their basic chemical structure (C6-C3-C6 skeleton) has a fused ring system consisting of an aromatic ring and an oxygen-containing heterocyclic benzopyran ring with a phenyl substituent. Flavonoids can be divided into different classes depending on their oxidative status and substituents. In nature, flavonoids often occur as polymers, with dimers being the most common form. Most flavonoids, apart from cathecins, are usually present in plants as β-glycosides. (Fig. 2.8).

The flavonoids appear to have an important role in the successful medical treatments of the ancient times, and since the last century they are subject of increasing interest of scientists working in various fields of medical research, including pharmacology of parasitic diseases. The excellent review on biochemistry and medical significance of the flavonoids was written by Havsteen (2002). His review deals also with a high spectrum of positive effects of various flavonoids on plant and mammalian cells in relation to therapeutical applications, their ability to inhibit specific enzymes, to simulate some hormones and neurotransmitters, and to scavenge free radicals. It is well known that some flavonoids can inhibit or kill many bacterial strains, inhibit important viral enzymes, such as reverse transcriptase and protease, and destroy some pathogenic protozoans.

In the many plant extracts which showed a high anthelmintic activity, chemical analyses revealed the presence of flavonoids, along with other classes of phytochemicals. Although toxicity of most isolated flavonoids to animal cells is very low (Middleton et al. 2000), several ubiquitous flavonoids genistein, kaemferol, rutin, quercetin, etc,. showed deleterious effects on selected species of parasitic helminths. It is possible, that different developmental stages (larvae, juvenile or adult) of helminths might possess different susceptibility to selected flavonoids. For example the phenolic diketone curcumin and the flavonoid kaempferol exerted a strong adulticidal effect on *Schistosoma mansoni,* but no activity against nematodes was demonstrated. *Kaempferol* (flavonol) and its three derivatives, were isolated from two plant species of *Styrax camporum* and *Styrax pohlii* (Braguine et al. 2012), and were examined in vitro. Of these, kaempferol was the most effective in separating *S. mansoni* male and female couples and killing adult worms at a concentration of 100 μg/ml. The selective toxicity of the flavones *quercetin, chrysin, and 3-hydroxyflavone* toward the several cancer cell lines but not to normal mammalian cells is worthy of mention (Pilatova et al. 2010). Selected flavones stimulated mechanisms leading to apoptosis of cancer cells and inhibited functions of important cell signaling molecules.

The polyphenol *curcumin* (phenolic diketone) is the major constituent in the rhizome of *Curcuma longa* (Zingiberaceae), which is responsible for the characteristic yellow pigment. Curcumin is well known to exhibit several biological activities, including anti-inflammatory, antioxidant, antiviral, anti-infectious, and anti-carcinogenic activities (for example Maheshwari et al. 2006). Angiogenesis is a key step in tumor growth and invasion and, in the recent review of Varinska et al. (2010), the anti-angiogenic potential of polyphenols including curcumin was discussed. Curcumin was shown to cause the death of all worms of *S. mansoni* at 50 and 100 μM concentrations due to the decreasing of their viability (Magalhães et al. 2009). All pairs of coupled adult worms were separated into individual male and female by the action of curcumin at the doses of 20–100 μM and it also reduced egg production by 50 %. An important issue in the drug development process is drug solubility in water and pharmacokinetic properties in vivo, which can be manipulated after their incorporation into drug delivery systems (DDS). The DDS should deliver a biologically active molecule at a desired rate for a desired duration and at a desired target, so as to maintain the drug level in the body at optimum therapeutic concentrations with minimum fluctuation. In the recent study of Luz et al. (2012) curcumin was incorporated into poly (lacticco-glycolic) acid (PLGA) nanospheres by the nanoprecipitation technique. Incubation of adult *Schistosoma mansoni* with curcumin-loaded PLGA nanoparticles caused the death of all worms, a separation between 50 and 100 % of worm couples and the partial alterations in the tegument at concentrations from 30 μM. Nanoparticles contributed to the higher absorption rate of entrapped curcumin in comparison with free compound, thus decreasing the threshold concentration from 100 to 30 μM. Interestingly, in vivo studies with curcumin showed a lack of significant toxicity (Perkins et al. 2002). The mechanism by which curcumin exerts its in vitro schistosomicidal effect is not clear. However, it has been reported that it has a direct action involving parasite biochemical processes. One of the possible targets in *schistosomes* for the curcumin action is the ubiquitin–proteasome pathway. In the study of Allam (2009), a total dose of 400 mg/kg BW of curcumin was given to mice with *S. mansoni* infection (in 16 injections) and treatment was effective in reducing worm burden by 44.4 % and tissue-egg burden by 30.9 %. Moreover, modulation of cellular and humoral immune responses was observed as well as a decrease of hepatic granuloma volume and overall collagenesis.

Genistein (4′,5,7-hydroxyisoflavone), a major component of soya, is a well-known phytoestrogen, which was found also in the ethanol extracts of higher plants *Flemingia vestita, Accacia* spp., *Stephania glabra,* and possibly others. The extensive studies have indicated beneficial effect of genistein on a multitude of disorders, including cancer, cardiovascular diseases, osteoporosis, and postmeno-pausal symptoms (see for review Barnes 1998). It has strong anti-inflammatory and antibacterial properties in vitro and in vivo (for example, Verdrengh et al. 2004), however, in mammals frequent administration of the higher doses either as therapeutical or dietary supply might cause undesirable side effects (Klein and King 2007).

The anthelmintic activity of genistein or genistein-rich extracts from root of plant *Flemingia vestita* has been proven on the cestodes *Raillietina echinobothrida, Echinococcus multilocularis* and *Echinococcus granulosus* and the trematode, *Fasciolopsis buski* in vitro. *Flemingia vestita* is an indigenous medical plant of north-east India and, according to the ethnomedical experiences, the crude extract of the root-tuber peel was effective against trematodes and cestodes, but not nematodes. The high concentration of *isoflavones* was found in the extract from this plant (Rao and Reddy 1991). The anthelmintic activity of genistein is mediated by its action of several cellular/molecular targets, and such a complex mode of action of this flavonoid was also found for mammalian cells. In parasitic flatworms, the primary or vital targets of *genistein* were suggested to be the tegumental enzymes, and their inhibition probably resulted in the paralysis and worm death (Pal and Tandon 1998). In vitro exposure to this compound resulted in tetanic contractions, flaccid paralysis, and disruption of tegument in *R. echinobothrida* (Tandon et al. 1997). After incubation of worms with genistein (0.5 mg/ml) activity of enzymes AcPase, AlkPase, ATPase, and 5′-Nu was found to be suppressed by 97, 95, 88, and 57 %, respectively. The effect of genistein on a neuromuscular signaling pathway in worms is implicated by results of another study (Kar et al. 2002). Nitric oxide (NO) is the important neuronal messenger, and the enzyme nitric oxide synthase (NOS) catalyzes the conversion of L-arginine to citrulline and nitric oxide (NO). The enzyme NOS exists in three isoforms, which are either constitutively expressed in endothelial cells (cNOS), neurons (nNOS) or are induced by endotoxin and the inflammatory cytokines (iNOS). The activity of the first two NOS depends on the intracellular concentration of Ca^{2+}. In the trematode *Fasciolopsis bruski*, which is the large intestinal fluke of swine and humans, genistein treatment in vitro increased nNOS activity in the neuronal tissues and consequently, elevated production of NO, which might also account for, among other factors, onset of paralysis—a mani-festation of neurotoxicity. In helminths, NO has been suggested to have many physiological roles as a neurotransmitter at neuromuscular junctions, and in control of embryogenesis (Pfarr et al. 2001).

At the molecular level, NO exerts its most relevant physiological action by activating the soluble form of guanylyl cyclase, leading to the accumulation of

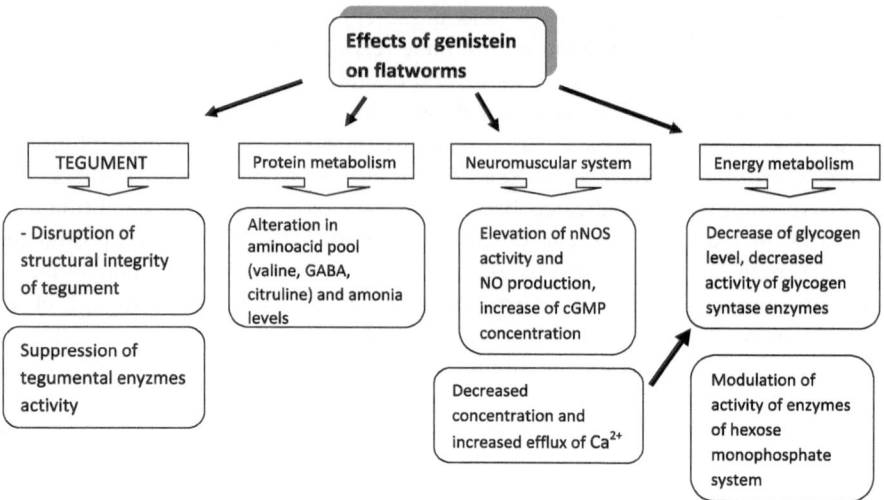

Fig. 2.9 Multitargeted effect of genistein in flatworms

cGMP, an important messenger mediating the functions of NO inside the cells. In the cestode *R. echinobothrida*, incubation of worms with genistein (0.5 mg/ml) modulated this physiological pathway in parasitic tissue (Das et al. 2009). At the time of onset of paralysis in the parasites, a significant increase (32–87 %) in the NOS activity, a two to three fold increase of NO efflux into the incubation media as well as the elevation of cGMP concentration in the treated parasite tissues by 44–103 % were observed. Data indicate that genistein can disturb the downstream signaling pathway of NO, as indicated by the change in cGMP concentrations in parasites. Calcium, which is stored in the calcareous corpuscles of many cestodes, especially the larval (metacestode) stages, is intimately involved in both muscle contractions and signal transduction of many receptors. In *R. echinobothrida* following in vitro treatment with genistein, the Ca^{2+} concentration was decreased significantly by 39–49 % in parasite tissue and also an increase of Ca^{2+} efflux by 91–160 % into the culture medium. The changes in Ca^{2+} homeostasis may be related to the rapid muscular contractions and consequent paralysis in the parasite due to the anthelmintic stress caused by the phytochemicals (Das et al. 2006). A similar effect on breakdown of Ca^{2+} homeostasis in the tegument of flatworms is known for praziquantel (Cioli and Pica-Mattoccia 2003). It seems that neuro-muscular activity of genistein is one of several effects, indicating multiple targets in parasitic cells which are summarized in Fig. 2.9. It was shown that genistein can interfere with the energy metabolism of flatworms. After exposure to genistein at a concentration of 0.2 mg/ml, the glycogen concentration in *R. echinobothrida* decreased by 15–44 %, which was accompanied by increase of activity of the active form of glycogen phosphorylase by 29–39 % and decrease of activity of the active form of glycogen synthase by 36–59 %. (Tandon et al. 2003). PZQ (1 μg/ml) the reference drug, also caused quantitative reduction in glycogen level and alteration in enzyme activities. With limited ability to metabolize lipids and amino acids, cestodes and trematodes mainly utilize glucose and other simple carbohydrate molecules to meet their energy requirements (Bryant and Behm 1989) and the alterations in the activity of enzymes regulating glycogen metabolism may also be related to Ca^{2+} efflux. Several other enzymes are involved in energy metabolism of flatworms and genistein significantly influenced the key enzyme of hexose monophosphate pathway: glucose 6-phosphatase dehydrogenase and also enzymes of gluconeogenesis: pyruvate carboxylase, phosphoenolpyruvate carboxykinase, and fructose 1,6-bisphosphatase in *R. echinobothridia* in vitro (Das et al. 2004), which is perhaps a function of high energy demand of the parasite under anthelmintic stress.

Proteosynthesis and proteolysis in the worms are highly regulated and interconnected processes and genistein was able to induce the alterations in the free amino acid pool and ammonia levels in the fluke *Fasciolipsis buski* (Kar et al. 2004). Valine was found to be the most elevated amino acid as well as the levels of GABA and citrulline, which could be associated with the elevated activity of nNOS. Ammonia in the tissue homogenates as well as in incubation medium increased (66.4 %) compared to the controls. These data might indicate that genistein could be an effective compound in the therapy of gastrointestinal cestode and trematode infections, but it is necessary to determine the effective concentration and treatment schedule on animal models. There is a risk that the higher doses might lead to the uncontrolled stimulation of selected immune mechanisms, which are known to be the targets of genistein in mammals. It was shown that genistein, kaempferol, and quercetin can inhibit activation of the signal transducer and activator of transcription 1 (STAT-1), important transcription factor for iNOS and NO production in activated macrophages (Hämäläinen et al. 2007). The estrogen-like activity of genistein is a major concern during long-term chemotherapy as it binds to estrogen receptors and can induce estrogenic effects. Prolonged treatment with current anthelmintics is required to inhibit growth of parasitic cysts of *E. multilocularis*. It was shown that genistein exhibited significant metacestodicidal activity against *E. multilocularis* in vitro as well as against *E. granulosus* metacestodes and protoscoleces (Nagulesvaran et al. 2006). The native compound and synthetic derivatives of genistein, Rm 6423, and Rm 6426 induced truncation of microtriches, nuclear pyknosis, and vesiculations of protoscoleces. In addition, Rm 6423 specifically induced dramatic breakdown of the structural integrity of the germinal layer in metacsetode cysts and a decrease of activity of metalloproteases, which allows the growth of cysts in the host tissues. Inter-individual, species and sex differences in gastrointestinal metabolism of this phytoestrogen may be critical factors in determining the efficacy of these various compounds in vivo.

There are a few studies, in which nematocidal activity of other plant-derived flavonoids was demonstrated in vitro and in vivo. The *antifilarial activity* of 6 flavonoids against the human lymphatic filarial parasite *B. malayi* was evaluated using an in vitro motility assay with adult worms and microfilariae, a biochemical test for viability (MTT-reduction assay), and two animal models, *Meriones unguiculatus* (implanted adult worms) and *Mastomys coucha* (natural infections)

(Lakshmi et al. 2010). All six flavonoids showed antifilarial activity in vitro, which can be classed in a decreasing order: *naringenin* > flavone = hesperetin > rutin > naringin > chrysin. IC_{50} of naringenin was 2.5 µg/ml and adulticidal effect was seen at 125 µg/ml. In jirds, naringenin and flavone killed or sterilized adult worms at dose of 50 mg/kg, but in *Mastomys*, where the parasite produces a patent infection, only naringenin was filaricidal. All the flavonoids tested were well tolerated in both the animal models and there were no signs of behavioral or other changes that can be related to flavonoid treatment in the animals. Killing the adult worms or sterilizing the female worms is considered to be one of the best strategies, however, the mechanism by which these flavonoids affect the viability of filarial parasites is unknown.

The methanol extract of *Struthiola argentea* whole plant exhibited in vitro activity on nematodes and was therefore selected for bioassay-guided fractionation using the in vitro *Haemonchus contortus* assay. The newly isolated methoxylated flavone, *flavone 3* exhibited the most potent activity with an EC_{90} of 3.1 µg/ml, which is significantly (>17-fold) less active than the ivermectin control (EC_{90} = 0.18 µg/ml). (EC = effective concentration). However, in vivo evaluation using the *Heligmosomoides polygyrus* mouse model revealed that flavone 3 did not have any in vivo activity at 25 mg/kg (Ayers et al. 2008). In the case when high in vitro and much lower in vivo activities are observed, the possibility should be taken into consideration that host metabolism might transform one class of flavonoid into another, resulting in generation of new pharmacological activity or the loss of previous activity.

2.3.5 Glycosides and Saponins

Saponins and glycosides are naturally occurring chemical compounds found in a wide variety of higher plants, for example in lucerne (*Medicago sativa*), sisal (*Agave sisalama*), and seeds like soyabeans (*Glycine max*) (Francis et al. 2002). In chemistry, glycoside is a molecule in which a sugar is bound to a non-carbohydrate moiety, usually small organic molecules. They play numerous important roles in living organisms. Many plants store chemicals in the form of inactive glycosides, which are activated by enzyme hydrolysis. Saponins are chemically *glycosides*, which are composed of a lipid-soluble aglycon consisting of either a sterol or more commonly a triterpenoid with different, water-soluble, sugar residues. Saponins also have surfactant properties and sterol-group-containing saponins can particularly affect eukaryotic organisms that contain steroids in their membrane. They also demonstrate hemolytic action toward red blood cells, and can be toxic if given intravenously (Osbourn et al. 2011). Moreover, consumption of food with a high saponin concentration by humans and animals can be potentially harmful (Milgate and Roberts 1995). Glycosides and saponins were detected in many higher plants used in the traditional ethnomedicine at various concentrations. So far, only several of these molecules isolated from the plants have been examined for their anthelmintic activity under the experimental conditions or in vivo controlled studies with promising results (Makkar et al. 2007).

In vivo study conducted on rats with *H. diminuta* infection, *saponins* extracted from *Acacia auriculiformis* were administered orally at the dose of 200 mg/kg for several days. Adult worms were expelled within several days from intestinum indicating that saponins induced tegumental damage followed by the worm paralysis, which usually precedes expulsion. At such relatively higher dose of crude extracts no side effects on the host were observed (Ghosh et al. 1996). Chemical analysis of extract from the funicles of *A. auriculiformis* revealed the presence of triterpenoid *saponins acaciaside A and B*. They are the unique molecules because they contain a conjugated unsaturated system, which is highly susceptible to peroxidation (Ghosh et al. 1993). Except of cestocidal effect, these saponins showed also strong antifilarial activity, when concentration of 4 mg/ml killed 97 % of microfilaria of *Setaria cervi* and 100 % of adults within 100 min (Ghosh et al. 1993). Significant reduction of infection was found in vivo after repeated doses of 100 mg/kg without seeing any toxic effects in rats. Drugs probably caused a very high physiological stress on adult worms, resulting in their death and expulsion which is opposite to the low efficacy of antifilarial drugs toward adults. Sinha Babu et al. (1997) showed experimentally that these saponins enhance the cell membrane lipid peroxidation. They and others (Nandi et al. 2004) suggested that the conjugated unsaturated system of selected saponins is involved in the formations of free radicals, which induce membrane damage through peroxidation of membranes in helminths.

Glycosides with strong antifilarial activity were isolated from the crude extracts of the stem bark of *Streblus asper*, traditional medical plant of India. Two cardiac glycosides *asperoside* and *strebloside* were highly effective at the dose of 50 mg/kg against *Brugia malayi* in vivo and also in vitro, however several cardiac glycosides of other origins did not show any comparable antifilarial efficacy (Chatterjee et al. 1992). Different types of saponins were found in a high concentration in sisal *(Agave sisalama)*, which is an important crop in Brazil. About 60 % of plant material is composed of liquid, containing mostly *sapogenins*, the non-glycosidic portion of saponins, which demonstrated significant anthelmintic activity on L1 stage of gastrointestinal nematodes (Silveira et al. 2012). Three types of sapogenins were detected in sisal, of which *hecogenin* was found of the greatest quantity. Saponins also have surfactant properties and can particularly affect eukaryotic organisms that contain steroids in their membrane (Osbourn 1996). Steroidal saponins are considered to be the active ingredients of plant extracts of sisal, which have detergent properties similar to those of polyene antibiotics. According to Francis et al. (2002), saponins present in *A. sisalana* caused their effect by intercalation in the cell membranes by their hydrophobic fraction, causing the formation of pores in tegument of helminths. Other steroidal saponins were isolated from the methanol extract of *Dracaena fragrans* (Agavaceae) and were tested on adult worms of trematode *Schistosoma mansoni* in vitro.

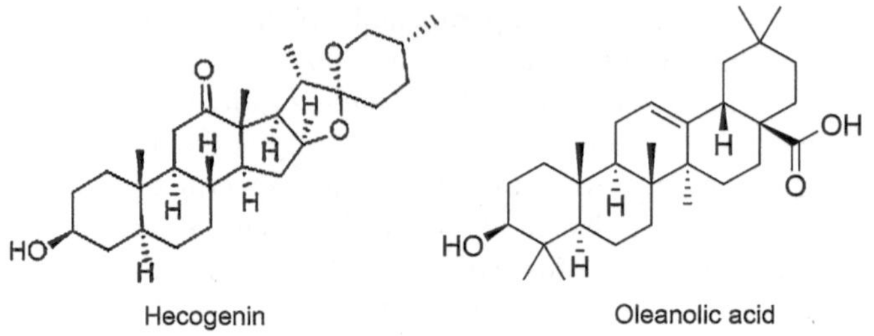

Hecogenin   Oleanolic acid

Fig. 2.10 Chemical structure of selected components of saponins with anthelmintic activity

Lethal effect on adults was achieved with LC_{50} of 18.4 μg/ml 4 days after exposure (Tadros et al. 2008).

Calendula officinalis and *Beta vulgaris* are plants native in countries with temperate climate and their saponin components were detected as *triterpenic pentacyclic oleanolic acid* (OA, oleanane-type triterpene) *glucosides with glucuronic acid* attached to the hydroxyl group at the C-3 position of aglycone (oleanolic acid) (Fig. 2.10).

The glycoside compounds connected to glucuronic acid differ in saponins of both plant species (Doligalska et al. 2011). The anthelmintic activity of oleane-type glucuronides (GlcUAOA) was examined on the development of free-living stages of *Heligmosomoides bakeri*, a parasitic nematode of the mouse intestine. Both *C. officinals* and *B. vulgaris* GlcUAOA affected the development of the free living stages and interfered with function of the major membrane transporter for xenobiotics, P-glycoprotein (Pgp) in *H. bakeri*. The GlcUAOA inhibited egg hatching and molting of larvae and also changed their morphology. In nematodes, the availability of drugs is modulated by physical and biochemical barriers such as cuticle and intestinal epithelium (Kennedy et al. 1987; Kerboeuf et al. 2010). The function of these barriers depends on specific membrane transport systems such as P-glycoprotein, and nematode resistance to anti-parasitic treatment may be mediated by Pgp-related pathways (Kerboeuf et al. 2003, Kerboeuf and Guegnard 2011). Pgp plays a crucial role in the distribution, metabolism, excretion and absorption of toxic molecules (Riou et al. 2010). The mechanism of action of these saponins is not yet understood, but the anthelmintic activity could be attributed to the molecular structure of GlcUAOA, which is based on a 30-carbon skeleton comprising 5 six-membered rings (*ursanes and oleananes*). Oleanolic acid, when glycosylated at both C-3 and C-28, induces a permeability change in the cell membranes (Hu et al. 1996) probably also in cuticles and cell membranes of *H. bakeri* larvae (Doligalska et al. 2011). The integrity of cell membrane is critical for the barrier function and its loss results in cell death. Interestingly, the increasing level of Pgp has been observed in nematode strains resistent to anthelmintic treatment (Kerboeuf et al. 2003).

2.3.6 Enzymes, Amides and Other Specific Compounds

A broad group of plants, which has been used traditionally for the treatment of helminth infections, includes papaya (*Carica papaya*), fig (*Ficus* spp.), and pineapple (*Ananas comosus*). Papaya and fig release latex upon injury, which is rich in proteolytic enzymes, whereas other plants, such as the pineapple, contain large amounts of cysteine proteinases in the juices extracted from the stems or fruits (Rowan et al. 1990). These enzymes are already in use in medicine for their beneficial effects for example, during inflammatory diseases, and, interestingly, they were shown as the active anthelmintic principle toward nematodes (see reviews: Behnke et al. 2008; Stepek et al. 2004). *Cysteine proteinases* (CPs) present in the fiber lattices and extracts of fruits, all have a neutral pH optimum of around 7 and the enzymatic activity is associated with the soluble fraction after centrifugation. All of these cysteine proteinases from plants have similar, but not identical activities, and they vary in other important characteristics such as resistance to acidic conditions and susceptibility to digestion by the enzymes of the alimentary tract. The primary site of their proteolytic activity in nematodes is the cuticule but, surprisingly, free-living and soil-dwelling stages of parasitic nematodes, as well as totally free-living species, are resistant to their action (Behnke et al. 2008). The different composition of the free-living nematode outermost cuticular layers and the presence of biochemical defenses—proteinase inhibitors, is the possible explanation. In the case of parasitic stages of nematodes living in the gut, in the absence of host secreted intestinal CPs, it was unreasonable for parasitic stages to develop the defences against any CPs (Zang and Maizels 2001).

Nematocidal activity of enzymes present in papaya latex (*papain, chymopapain, caricain,* and *glycyl endopeptidase*) was demonstrated on *Ascaris suum* in naturally infected pigs (Satrija et al. 1994), on *Heligmosoides polygyrus* (syn. *Heligmosomoides bakeri*), *Trichuris muris* and *Protospirura muricola* infections in mice (Satrija et al. 1995; Stepek et al. 2006; 2007a, b, c; Behnke et al. 2008), all in monogastric animals. Experimental *H. bakeri* infection in mice is used as model nematode infection for monogastric animals and humans. A daily administration of papaya latex to infected mice for 7 days (133 nmol active cysteine proteinase/mouse/day) resulted in a nearly complete elimination of worms by day 25 post-therapy, indicating by 97 % reduction of *H. bakeri* fecal egg counts (Behnke et al. 2008). Recently, Buttle et al. (2011) showed that enzymes present in papaya latex posses potent anthelmintic activity capable of clearing the adult parasitic nematode *Haemonchus contortus* from the sheep abomasum. The lack of efficacy of a single dose compared with the use of 4 daily doses suggests that, following dilution in the rumen, the enzymes require prolonged contact time with the worms in order to prove effectiveity.

Treatment with latex from the South American fig (*Ficus glabrata)* containing the enzymes *ficin and ficain*, was evaluated in preclinical study on groups of residents infected with one or more of gastro-intestinal nematodes *Ascaris, Ancylostoma,* and/or *Necator, Trichuris* or *Strongyloides* (Hansson et al. 1986). A dose of 1 ml/kg significantly reduced eggs per gram (EPG) for all of these nematodes but complete elimination of worms was not achieved. Latex from other *Ficus* species, given at the dose of 4 ml/kg for 3 days, was examined as a vermifuge in mice naturally infected with several gastro-intestinal nematodes (de Amorin et al. 1999). The weak anthelmintic efficacy between 2.6 % up to 41 % as well as a high acute toxicity in gut, exclude ficin, the main principle of latex from fig, from therapeutical purposes. The enzymatic activity in *Ananas comosus*—pineapple is attributed to *ananain, fruit bromelain, stem bromelain, and comosain*, but their anthelmintic activity has been little examined in controlled in vivo experiments. Regarding the side effects of CPs, the immunogenic properties of orally administered fruit-derived CPs have only been examined by Hale (2004), who detected relatively low levels of circulating bromelain specific IgG after 18 weeks of daily oral treatment with bromelain, however, papain is known to be allergenic when inhaled.

Amides are small organic substances containing nitrogen in their molecule, similar to alkaloids. Amides are found in plants in much lower concentration than other secondary metabolites, where they usually play role in the defense against insects or fungi. The anthelmintic activity on *Schistosoma mansoni* was reported for amide *piplartine*, isolated from roots of plant *Piper tuberculatum* (de Moraes et al. 2011; de Moraes et al. 2012). The genus *Piper* includes species that are widely distributed throughout the tropical and subtropical regions of the world. Piplartine, 5,6-dihydro-1-[1-oxo-3-(3,4,5-trimethoxyphenyl)-2-propenyl]-2(1H)-pyridinone, is found in several *Piper* species and has shown several biological activities on humans as well as antifungal, insecticide (Navickiene et al. 2000, 2003) and antiprotozoan activities against visceral leishmaniasis (Bodiwala et al. 2007). Piplartine concentration of 15.8 μM reduced the motor activity of adults *Schistosoma* worms and caused their death within 24 h in vitro, probably also due to extensive tegumental destruction and damage to the tubercles. A concentration of 6.3 μM caused a 75 % reduction in egg production, however, separation of worm couples was not observed. It is known that the larval stage of this trematode, the schistosomulum, which infects humans via skin, is not sensitive to praziquantel therapy. In the recent study of de Moraes et al. (2012) piplartine reveals interesting anti-schistosomal properties also on schistosomula of different ages (3 h old and 1, 3, 5, and 7 days old). An extensive tegumental destruction, including blebbing, granularity and a shorter body length was observed, followed by the death of parasites, after 120 h of incubations with 7.5 μM of this amide. The mechanism by which piplartine exerts its in vitro schistosomicidal effects is not clear and, according to de Moraes et al. (2011), in vitro antischistosomal effects of piplartine may be related to the inhibition of neurotransmission system pathway in *S. mansoni*. They observed that in vitro effects of piplartine on *S. mansoni* adults better correlated with the muscular function (motor activity) than with tegumental destruction.

The rhizomes of fern *Dryopteris* spp. have popularly been used as vermifuge in flatworm infections, where the active anthelmintic principles were considered to be *phloroglucinol derivates aspidin, flavaspidic acid, methylene-bisaspidinol*, and *desaspidin* (Socolsky et al. 2009). The main effects were tegumental alterations, decrease of motor activity of adult worms, egg production and decreased development of eggs produced by the adult worms, which were inhibited by the incubation with these compounds in the concentrations range from 10 to 100 μM (Magalhães et al. 2010). These authors suggest that these effects may be related to the inhibition of oxidative phosphorylation pathway in *S. mansoni*.

Very promising compounds isolated from *Xylocarpus granatum* (local name in India is fruit from Andaman) with high antifilarial activity are *gedunin* and *photogedunin*. Of several isolated compounds, only two: gedunin (IC = inhibition concentration, CC = cytotoxicicity concentration determined on VERO cell lines, SI = selectivity index) (IC_{50} = 0.239 μg/ml, CC_{50} = 212.5 μg/ml, SI = 889.1) and photogedunin (IC_{50} = 0.213 μg/ml, CC_{50} = 262.3 μg/ml, SI = 1231.4) at five daily doses of 100 mg/kg given by subcutaneous route revealed excellent adulticidal efficacy resulting in the death of 80 and 70 % transplanted adult *B. malayi* in the peritoneal cavity of jirds (Misra et al. 2011). The high IC_{50} may indicate that these compounds blocked an essential metabolic pathway in the parasites and are worth of further examination (Fig. 2.11).

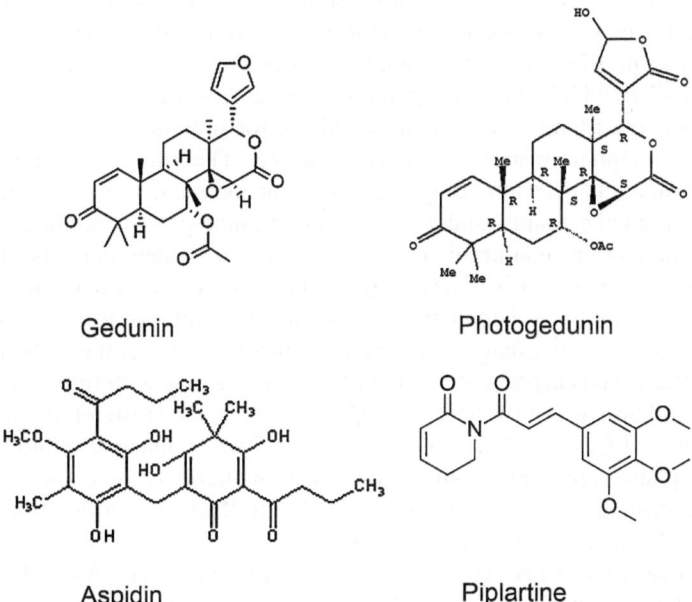

Gedunin

Photogedunin

Aspidin

Piplartine

Fig. 2.11 Chemical structure of molecules with marked anthelmintic activity: piplartine, aspidin, gedunin and photogedunin

2.3.7 Condensed Tannins and Sesquiterpene Lactones

Tannins are group of secondary plant metabolites formed by water-soluble phenolic compounds with a great diversity, which can be divided into two major

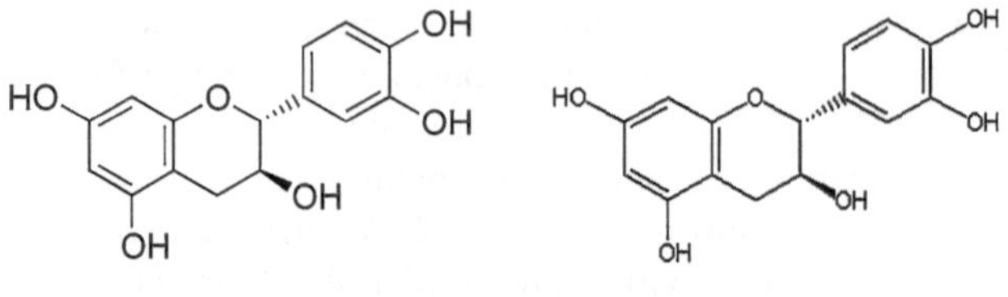

Catechin   Epicatechin

Fig. 2.12 Chemical structure of components of condensed tannins (flavan-3-ols) catechin and epicatechin

groups: the hydrolyzable and the condensed tannins (Waterman 1999). Condensed tannins are *polyphenols (=proanthocyanidins)* with high molecular weight from 500 to 20,000 kDa, which consist mainly of oligomers or polymers of monomeric units of flavan-3-ols (catechin, epicatechin, and so on) (Fig. 2.12). Depending on the chemical structure of the monomeric unit, in particular the number of hydroxyl radicals, they are classified into four sub-classes: the prodelphinidins (PDs), procyanidins (PCs), prorobetinidins (PRs), and profisetinidins (PSs). In legume forages, PCs and PDs are mainly found, in various proportions. The ratio between PDs/PCs differs widely between plant species and/or varieties (Mueller-Harvey 2006). Condensed tannins have a high affinity for proteins and polysaccharides, which can precipitate from the aqueous solutions. The affinity of condensed tannins for proteins is determined by the molecular mass and the molecular configuration of both the tannin and the proteins. Tannin–protein binding is usually reversible in acid or alkaline pH or after treatment with detergents (surfactants).

Condensed tannins (CTs) have high relevance for livestock production as tannin–rich plants have a long tradition of use not only as forages but also as "green" control of gastrointestinal nematode infections. Several excellent reviews deal with the various aspects of feeding of small ruminants with forages containing tannin-rich plants or even fodder trees (Diaz et al. 2010; Hoste et al. 2005, 2006, 2012; Sandoval-Castro et al. 2012 and others). They pointed that bioactive tanniniferous plants represent a valuable option as an alternative to commercial drugs for the control of gastro-intestinal nematodes (GINs) as consumption of these plants has been associated with antiparasitic effects: reductions in nematode numbers, worm fecundity, and nematode eggs excretion. The main threat to the use of solely chemical drugs is the rapid development of resistance to any anthelmintic drug in worm populations after commercialization (Waller 2006) and the spread of anthelmintic resistance within worm populations (Kaplan 2004). These problems have stimulated research on plant-based phytochemicals with anthelmintic activity against gastrointestinal nematodes. Although consumption of high concentrations of condensed tannins (>7 % of DM) had a number of detrimental effects on ruminants, such as reduction in food intake, growth inhibition and interference with the morphology and the proteolytic activity of microbes in the rumen (Min et al. 2003; Waghorn and McNabb 2003), low or moderate concentrations of condensed tannins (<6 % of DM) have resulted in the positive effects on animals. The pertinent use of *tannin-rich fodders* as *nutraceuticals* supposes a clear understanding of the mode of action against the worms. The term "nutraceutical" results from a contraction of nutrition and pharmaceutical. It is defined as "any substance that may be considered a food or part of a food which provides health benefits, including the prevention and treatment of disease" (Andlauer and Furst 2002).

Several aspects of direct and indirect effects of bioactive tannin-rich tropical and temperate legumes against nematode infections have been analyzed in detail in the recent review of Hoste et al. (2012). They discussed high variability of the effects of plant extracts on GIN and pointed that a way to overcome the origins of such variability is to better understand the mode of action of the bioactive compounds against the various nematode stages. This means: (i) to discuss the nature of the secondary metabolites involved in the activity; (ii) to better understand how these compounds affect the different parasitic stages and their biological or functional traits; and (iii) to relate these changes to possible consequences on the parasites' life cycles.

The other group of the biochemical compounds with anthelmintic activity found in plants used as forage, for example in chicory, are *sesquiterpene lactones* (Foster et al. 2006). Within the last decade a number of studies focused on isolation of condensed tannins and sesquiterpene lactones from various legume forages and plants with the aim to reveal their effects in vitro and in vivo on various species and developmental stages of nematodes. Differentiated action of *condensed tannins* on parasite stages was observed by Athanasiadou et al. (2001), which were more effective against larvae than adults. This can also be explained by the difference between the cuticular components of the pre-parasitic stages (eggs to L3) and the parasitic stages (L4 and adults), as demonstrated by the study of Stepek et al. (Stepek et al. 2007a, b). CTs do have significant negative effects upon egg hatching and larval hatching in vitro as was demonstrated for example by Molan et al. (2002) and Novobilský et al. (2011).

Forage CTs are polymers of flavan-3-ol units, with a considerable range of structural variations. The constituent flavan-3-ol units in procyanidin (PC) polymers are either catechin (C) or epicatechin (EC), while prodelphinidin (PD) polymers contain either gallocatechin (GC) or epigallocatechin (EGC). *Trichostrongylus colubriformis* infective larvae were exposed to flavan-3-ols and their galloyl derivatives under in vitro conditions to compare their effects on the viability of eggs, development of first stage (L1) larvae, and the viability of the infective larvae determined by their mobility (Molan et al. 2003a). The flavan-3-ol gallates were effective in all three tests, with epigallocatechin gallate being the most effective in the egg hatch test (100 % inhibition at 1 mg/ml), also in inhibition of viability of larvae at 500 μg/ml. There was complete inhibition of development by all compounds at 100–200 μg/ml and 50 % inhibition between 42 and 59 μg/ml for flavan-3-ols, while values between 32 and 48 μg/ml were observed for flavan-3-ol gallates. Flavan-3-ols caused some inhibition of viability, but were not effective on other developmental stages (Molan et al. 2003a). The active CT extracts from forage legumes have epigallocatechin as the dominant flavan-3-ol extender unit, and epigallocatechin was the most active flavan-3-ol in both the EHT and LDA assays. Results may indicate that the anti-parasite properties of CTs are not significantly dependent on their structure in terms of 2,3-stereochemistry of the heterocylic C-rings (2,3-cis or 2,3-trans), but rather on the number of hydroxy groups in the B-ring (PD:PC ratio). The degree of polymerization and the inter-flavanoid linkages may play a significant role in the effects of CTs. *Onobrychis vicifolia* (sainfoin) is other leguminous forage rich in condensed tannins, which significantly inhibited migration ability of *H. contortus* L3 stage of larvae in vitro (Barrau et al. 2005). Inhibition was seen also for other bioactive compounds flavonol glycosides (rutin, nicotiflorin, and narcissin) present in low molecular weight fraction (up to 2 kDa) implying that both classes of phytochemicals contributed to the anthelmintic activity of this plant in vivo.

Chicory (*Cichorium intybus*) is a widespread herb with high nutritional value, that contains several secondary compounds, with sesquiterpene lactones and the major phenolics (condensed tannins) being the most abundant. The levels of the sesquiterpene lactones (*lactucin, lactupicrin,* and *8-deoxylactucin*) and the *hydroxyl coumarin chicorin* were found to be highest in the most actively growing regions of the plant (Rees and Harborne 1985). CT and crude sesquiterpene lactones (CSL) extracted from chicory have been shown to have direct effects on the motility of first-stage (L1) and third-stage (L3) larvae of deer lungworm (*Dictyocaulus viviparus*) and L3 larvae of gastrointestinal nematodes in vitro using the larval migration inhibition (LMI) assay (Molan et al. 2003b). Condensed tannins appeared to be more effective than CSLs at inactivating L1 and L3 lungworm and L3 gastrointestinal larvae in rumen fluid, but CSLs were particularly effective against L3 lungworm larvae in abomasal fluid. Condensed tannins were also effective at reducing the motility of infective larvae of gastrointestinal nematodes. Moreover, the L3 larvae of gastrointestinal nematodes were more sensitive to CT than L1 and L3 larvae of lungworms. This may be attributed to the fact that gastrointestinal L3 larvae were exsheathed, while the lungworm larvae were not. The protein surface of the sheath may interact with the CT and protect the larvae, whilst the absence of a sheath brings the larvae into direct contact with the CT and thus exposes them to a greater paralyzing effect. This was confirmed in the study of Molan et al. (2000b) who found that exsheathed larvae of deer gastrointestinal nematodes were more susceptible to the actions of CT than larvae with protecting layers. Indeed, Brunet and Hoste (2006) showed that monomers of condensed tannins affect directly the larval exsheathment of parasitic nematodes of ruminants. CSLs were effective at reducing the motility of lungworm and gastrointestinal nematode larvae, but their inhibitory activity in rumen fluid was lower than found for CT. This finding may be of practical value in controlling infection of the abomasum and intestine. In contrast to CT, pH probably does not affect the reactivity of CSL. The addition of PEG did not affect the biological activity of CSL, which may indicate that CSLs inhibit these larvae by a different mechanism than do CT.

The proportions of the *sesquiterpene lactones* (lactucin = LAC, lactupicrin = LPIC and 8-deoxylactucin = DOL) vary among forage chicory cultivars, that could modify the final anthelmintic activity of CSL (Foster et al. 2011). After using the combined LAC, DOL, and LPIC concentrations in the range from 0 to 5.0 mg/ml, egg hatching of *Haemonchus contortus* decreased sharply in a linear fashion. Concentrations of sesquiterpene lactones required for 50 % lethality were 2.6 mg/ml for cultivar containing higher amount of DOL and 6.4 mg/ml for cultivar with higher concentration of LAC, suggesting that LAC has minimal effect on egg hatching and that DOL or other constituent(s) in the CSLs are inhibitory (Foster et al. 2011). Metabolism or breakdown of free and bound sesquiterpene lactones in the rumen of animals is likely, and degradation rates of the individual chemical structures may vary, influencing what compounds and amounts reach the relevant section of the gastrointestinal track. Further studies are needed to determine the extent to which sesquiterpene lactone glycosides might influence the anthelmintic potential of chicory herbage.

Variations in the concentration and biochemical composition of condensed tannins extracted from several legume forages (*Lotus* spp., *Hedysarum coronarium*, *Onobrychis viciifolia*, *Dorycnium* spp.) were probably directly responsible for altered behavior of *Trichostrongylus columbriformis* exsheathed L3 stage larvae in a migration inhibition assay (Molan et al. 2000a, c). At 100 µg/ml, purified CTs from *Dorycnium* spp. were more effective in larval inhibition than CT from other forages. Similarly, at concentration of 1,000 µg/ml CT from *Dorycnium* spp. had the highest inhibitory activity (63 %) also on L1 stage of larvae, followed by CT from *Onobrychis viciifolia* (59 %). Data from selected studies briefly indicate the complexity of actions of condensed tannins from various sources on different developmental stages of GIN. The specific composition of CT and the stage-specific protein antigenic array of nematodes, which interact together may explain the observed variability. It is possible that CT can interfere with proteins involved in glucose metabolism as CT from *Chicorium intybus* inhibited glucose uptake in mammalian cells (Muthusamya et al. 2008).

2.3.8 Endoperoxide Sesquiterpene Lactone Artemisinin and Derivates

Plant extracts from the genus *Artemisia* (Asteraceae) have been used for long time in traditional Chinesse medicine for treatment of parasitic infections in humans and animals. Artemisia is now growing in many European countries and has also become naturalized in North America (Baraldi et al. 2008). Antiparasitic activity is attributed to the unique bioactive sesquiterpene lactone with an endoperoxide bridge, artemesinin, which was found in the high concentration in *Artemisia annua* (Fig. 2.13), *A. vulgaris,* and *A. absinthium.* A recent review reported that *A. annua,* the only commercial source of artemisinin, also contains over 40 flavonoids, some of which might potentiate artemisinin effect in extracts by inhibiting cytochrome P-450 enzymes, which degrade artememisinin (Ferreira et al. 2010).

Fig. 2.13 Immage of plant *Artemisia annua*, the primary source of natural artemisinin

   Artemisinin is considered as the most important plant secondary metabolite with respect to toxicity to a number of parasitic protozoan species (for review see: Muraleedharan and Avery 2009). The authors of these review summarized the progress in the development of anti-parasitic agents based on artemisinin and data from in vitro and preclinical studies on medically important protozoan and metazoan infections. Although effective, artemisinin suffers from drawbacks such as short plasma half-life, limited bioavailability and poor solubility, either in oil or water. On the basis of observations that the peroxide group in the molecule is responsible for its anti-parasitic action, initial efforts to improve these properties led to the development of first-generation artemisinin derivatives in the early 1980s. Of these, *dihydro-artemisinin*, was almost six times more potent in vitro than the parent compound on *Plasmodium* parasites in vivo (Janse et al. 1994). Within the next decade other semi-synthetic derivatives, *artemether*, *arteether*, and *artesunate* were prepared from the parent compound (Fig. 2.14). It is believed that their activity is released when the endoperoxide bridge is open, giving rise to peroxide molecules, which are the source of reactive oxygen species (Olliaro et al. 2001). Being recognized as an effective drug against malaria and other protozoan parasites of humans, the pharmacological potential of artemisinin and derivatives have been evaluated in a numerous studies employing several helminth species in

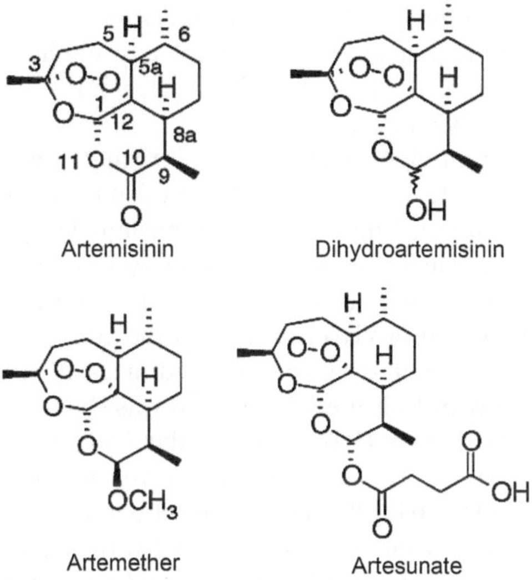

Fig. 2.14 Chemical structure
of artemisinin and derivates

Artemisinin   Dihydroartemisinin

Artemether   Artesunate

vitro and in vivo (see reviews: Utzinger et al. 2001, 2007; Keiser and Utzinger 2007; Xiao 2005a, b; Xiao et al. 2010). Thus these compounds, which were primarily explored as novel anthelmintics in the therapy of human parasitic infections, were recognized recently as the drug candidates for veterinary medicine. (Keiser et al. 2008; Squires et al. 2011).

The zoonotic disease fascioliasis, caused by the *liver flukes Fasciola hepatica* and *Fasciola gigantica* is of considerable public health relevance in Bolivia, Cuba, Egypt, the Islamic Republic of Iran, Peru and Vietnam. An estimated 91 million people are at risk and 2.4–17 million people are infected (Keiser and Utzinger 2005). Fascioliasis is also a problem for livestock production and a threat for global food safety (Schweizer et al. 2005). The only approved human anthelmintics for trematodiases are praziquantel and triclabendazole (TBZ), however the rapid spread of TBZ resistance is an important motivation for drug discovery of novel trematocidal drugs.

The effects of artemether and artesunate, derivates of artemisinin, were evaluated on adults and juvenile stages of both *Fasciola* species in the series of in vitro and in vivo studies. In vitro incubation in 10 µg/ml of artemether or artesunate caused severe tegumental damage on adult *F. hepatica* (Keiser and Morson 2008) comprising swelling of tegumental ridges, followed by blebbing and later rupturing of the blebs, leading to erosion and lesion, and disruption of the tegument. Similar alterations of the surface were also observed in adult *F. gigantica* when treated in vitro with 10–30 µg/ml of artemether (Shalaby et al. 2009) and in 3-weeks old juvenile stages after incubation with 20–80 µg/ml of artesunate for minimum 6 h. No fluke was found to be dead after incubation for 24 h (Tansatit et al. 2012).

An oral administration of artesunate at a single dose of 200 mg/kg resulted in 95 and 56.4 % reduction of worm burden in rats harboring adult and juvenile

Fasciola hepatica, respectively (Duthaler et al. 2010). A high oral dosage of 400 mg/kg artesunate completely eradicated adult *F. hepatica* in infected rats (Keiser et al. 2006a). Ultrastructural changes of the tegument and gut of the triclabendazole-resistant adult *F. hepatica* in the rats were observed following 24–72 h in vivo treatment with 200 mg/kg artemether (O'Neill et al. 2009). In all of these studies, it was concluded that the tegument of *F. hepatica* and *F. gigantica* is a primary drug target for artemisinin derivates and that they could be considered as drug candidates in the therapy of fascioliasis in monogastric animals.

Considering the veterinary importance of *Fasciola* spp. and the promising in vitro and in vivo activities of the artemisinins, trials were expanded to *ruminants* and their efficacy was assessed in sheep naturally infected with *F. hepatica*. In the study of Keiser et al. (2008) it was shown that single oral dose of 40 and 80 mg/kg body weight (BW) of artemether had no effect of FEC and worms burden, however a single dose of 160 mg/kg given intramuscularly significantly reduced the egg burden (64.9 %) and worm burden (91.3 %). A similar efficacy was achieved in naturally infected sheep treated with artesunate (Keiser et al. 2010). This indicates that the bioavailability after oral administration of artemether in sheep is much lower compared to drug bioavailability in monogastric animals that is related to the digestive physiology of the rumen. There were no adverse reactions observed in sheep after various single doses. The evaluation of the pharmacokinetic profile of both derivates under the same experimental design in the sheep model explained, at least partially, the ineffectiveness of the oral treatment (80 mg/kg) compared to intramuscular (i.m.) treatment (160 mg/kg). The observed plasma concentrations for derivatives and their metabolites were significantly higher following i.m. compared to oral treatment with an especially large difference observed for the metabolite dihydroartemisinin (DHA). This metabolite rapidly reduced the viability of worms in vitro, while other metabolites did not (Duthaler et al. 2012). In rats, artemisinins are primarily converted to DHA via ester hydrolysis and further to inactive metabolites by hepatic cytochrome P450 and other enzyme systems. DHA exhibits a low bioavailability after oral administration, with a short elimination half-life (Li et al. 1998).

The problem of high multiple doses of arthemisinins could be the embryotoxicity which was observed in laboratory animals (Clark et al. 2004). In addition, intramuscular administration of high doses of artemether has been shown to produce selective damage to the brain stem center of rodents, but only when multiple high doses of drug were administered. For example, intramuscular artemether (50–100 mg/kg day for 28 days) caused dose-dependent neuropathologic damage to the brain stem of mice (Nontprasert et al. 2002).

Schistosomiasis is a parasitic disease with a chronic debilitating character in humans and is ranked at second position of the world's parasitic diseases in terms of the extent of endemic areas and the number of infected people. Approximately 600 millions of people in 74 countries live at risk of infection (Lotfy 2009; Utzinger et al. 2009). Praziquantel is current drug of choice for the treatment of schistosomiasis. It is highly effective against adult stages and young developmental stages of all human schistosome species but is inactive on schistosomula

(Sabah et al. 1986; Xiao et al. 1987). There is a serious concern that a large scale application of praziquantel with having no similarly effective drug as an alternative, could lead to the development of tolerance/resistance to this drug. The first initiative in examination of artemisinins as anti-schistosomal compounds was taken by a group of Chinese scientists. They found that artemisinin administered to experimentally infected animals with *Schistosoma japonicum* resulted in marked reduction of adult worm burden (Chen et al. 1980). Since then numerous studies and reviews have been published dealing with in vitro and in vivo effects of artemisinin and its derivates on *Schistosoma species* (*S. mansoni, S. japonicum, S. mekongi* and *S. haematobium*). In the review of Utzinger et al. (2001) the potential of artemether for control of schistosomiasis was described in detail. Based on results obtained through extensive laboratory and clinical investigations it was concluded that oral artemether at a dose of 6 mg/kg administered in 2- or 3-weeks intervals is safe and is effective in the prevention of patent *S. japonicum* and *S. mansoni* infections, thus preventing the onset and evolution of pathology. Secondly, a combination therapy of praziquantel with arthemeter was proposed as effective alternative and this strategy has been recommended for effective transmission control. Tegumental changes induced by the semi-synthetic artemisinin derivatives artemether and artesunate have been assessed in different schistosome species, and were reviewed by Utzinger et al. (2007) showing a great similarity with *Fasciola* spp. In addition, they summarised data from preclinical studies and clinical trials showing the safety and efficacy in human. Only about 1 % of the participants with schistosomiasis and fascioliasis involved in the clinical trials reported mild abdominal pain, dizziness, headache, and diarrhea, or slight fever (Wu et al. 1995; Li et al. 2005; Keiser et al. 2011).

The extent of worm burden reduction following artemether therapy is dependent on the period of drug administration, since the drug selectively kills the larval migratory stage of parasite. Very high worm reductions (more than 90 %) were observed for 5–14 day-old schistosomula of *S. japonicum* (Xiao et al. 1995) and 14–21- day-old worms of *S. mansoni* (Xiao and Catto 1989; Xiao et al. 2000b). In the case of *S. hematobium*, which has the longest developmental period of 61–65 days until worms reach sexual maturity, the highest susceptibility to artemether was detected in 28-day-old schistosomula (Yang et al. 2001). In line with this, the highest efficacy (95–99 %) was achieved when the initial dose of artemether was administered to mice 3 weeks post infection (Xiao et al. 2000c).

It has been found in many studies that antiprotozoan activity is mediated by the endoperoxide bridge in the molecule of artemisinins. The mechanism of their action was subjected to the intensive research not only by parasitologists, but also immunologists as it was shown that these compounds have anti-cancer and immunomodulatory effects (Meshnick 2002; Shakir et al. 2011). Regarding their effect on trematodes, it was revealed that, in *S. mansoni*, sites of action of artemether include enzymes of glucose utilization. Worms recovered from artemether-treated hosts had an increased activity of glycogen catabolism and decreased glucose uptake (Xiao et al. 2000a). In mice infected with adult *Schistosoma japonicum*, the effect of artemether was evaluated on glutathione S-transferase

(GST) and superoxide dismutase (SOD), which are major antioxidant enzymes of schistosomes involved in detoxification processes (Callahan et al. 1988; Scott and McManus 2000a, b). In adult worms recovered from mice treated with either a subcurative (100 mg/kg) or a curative dose (300 mg/kg) of drug, significantly decreased activity of both enzymes was detected (Xiao et al. 2002). The higher inhibition was found for GST than for SOD, with female worms being more affected than males, resulting in about 55 % decrease of enzyme activity. It was suggested that this enzyme might increase the schistosome susceptibility to oxidative attack, and consequential to this, might be linked with the antischistosomal action of artemether.

The trematocidal effect of artemether and artesunate was confirmed also for *Clonorchis sinensis* and *Opistorchis viverrini*. In rats infected with 40–50 metacercariae of *C. sinensis* efficacy of treatment with a single oral dose of 150 mg/kg of artesunate, artemether was compared with that of praziquantel and tribendimidine. Reduction of adult worm burden was 100 % for artememisins and 89 and 80 %, respectively, for both anthelmintics. However, efficacy of the same dose of artemisinins against the juvenile stage of this trematode was considerable lower (57–59 %) in comparison with a high worm reduction with these drugs (Keiser et al. 2006b). Combination of PZQ or tribendimidine with either artesunate or artemether showed promising clonorchicidal properties (Xiao et al. 2008). The most effective combination was PZQ with artemether, at which many worms died due to extensive damage to tegument indicating the synergistic effect (Keiser and Vargas 2010). Worms collected from treated rats showed extensive damage to tegument already 8 h after artemether administration, including severe swelling, fusion, and vacuolization. Interestingly, the severity of tegumental changes did not progress further with time (Xiao et al. 2009). In rabbits infected with 300 metacercariae, the significant reduction of *C. sinensis* adult worm burden was achieved after oral administration of 120 mg/kg of artesunate (88.8 %) and artemether (67.2 %) (Kim et al. 2009). Rats are less suitable final hosts of *C. sinensis,* whereas rabbits are much more susceptible hosts to this fluke. Differences in the efficacy of two artemisinins in two different experimental models seem to be influenced by host factors, probably the physiology of rabbits and absorption/metabolisation rates. It is believed that a mechanism of action of artemisinins on *C. sinensis* and other liver flukes involves degradation of hemoglobin and generation of free heme, a possible target for peroxidic drugs (Utzinger et al. 2007).

Opisthorchiasis is a neglected tropical disease caused by the liver fluke *Opisthorchis viverrini* that affects the poorest people in Cambodia, Laos, north-eastern parts of Thailand, and Vietnam and praziquantel is the only available drug for this infection (Keiser and Utzinger 2005). Artesunate and artemether at a dose of 400 mg/kg given to hamsters infected with *O viverrini* resulted in worm-burden reductions of 77·6 % and 65·5 %, respectively (Keiser et al. 2006b), but when given to patients during randomized trial who had *O viverrini* infections, artesunate did not show any trematocidal effect (Soukhathammavong et al. 2011).

In contrast with a considerable interest in extracts from *Artemisia* spp. or pure artemisinins as trematocidal agents, to date a few studies examined their effects on other classes of parasitic helminths. In the recent study of Squires et al. (2011) artemisinin was tested for efficacy against *Haemonchus contortus* in a gerbil model of infection. Gerbils (*Meriones unguiculatus*) were recognized as a suitable model of *H. contortus* infection for anthelmintic testing, which offers the advantages of requiring smaller product quantities, greater standardization in testing conditions and evaluation of adverse effects on host pathology (Conder et al. 1990, Königová et al. 2008). Single oral doses of 400 mg/kg given on day 6 after infection or 200 mg/kg BW artemisinin administered daily for 5 days (between days 4 and 8 post infection) had no effect on reduction of pre-adult stages (L4) of this nematode. The lack of activity for L3 and L4 stages in gebrils, can not exclude that artemisinin and derivates, when given intramuscularly, will be effective on the adult stage.

The susceptibility of cestodes to treatment with artemisinin derivates was evaluated on the larval stage (protoscoleces) of the medically important cestodes *Echinococcus multilocularis* and *Echinococcus granulosus* (Spicher et al. 2008). As in the case of experimental infection with *H. contortus* in gerbils, no in vivo effect of a total dose of 200 mg/kg of artemisinin, artesunate, artemether or dihydroartemisinin was observed in infected and treated mice evaluated as the reduction in parasite cyst weight. Treatment began on week 8 post infection and a daily dose of 100 μl/mouse was applied by intragastric inoculation. This observation was in contrast with in vitro study, in which dihydroartemisinin and artesunate (10–40 μM) caused a 90 % reduction in viability of protoscoleces occurring on day 6 or 4 of incubation, respectively. Two other artemisinins were less effective. According to the authors, artemisinins and their metabolites were perhaps not delivered and accumulated in the parasite tissues in adequate quantities, because parasites are surrounded by an acellular laminated layer that represents a barrier for drugs. A short-elimination half-life of the active metabolite DHA probably contributed to its low bioavailability for echinococcus cysts (Li et al. 1998).

2.4 Concluding Remarks

Many people in developing countries still suffer or die from malaria, schistosomiasis, filariasis, and other infectious diseases, whereas, in developed countries, parasitic infectious diseases, especially those that are caused by opportunistic infection resulting from immunosuppressants and HIV/AIDS, are increasing. Moreover, the emergence of strains that are resistant to the current front-line drugs is reported nearly in all countries with intensive livestock production, emphasizing the need to search for compounds with antiparasitic activity for further drug development. Richness of bioactive molecules present in marine and terrestrial organisms and numerous records from traditional medicine and ethnoveterinary practice provide a starting point for pharmacological research. Essential parasite-specific systems, that differ from those of the hosts, represent attractive targets for specialized chemotherapy, as illustrated by glutamate-gated chloride channels and the specific activator avermectin. Isolation and chemical synthesis of sesquiterpene lactones with a unique endoperoxide bridge in their molecule—artemisinins, is another example of high potential of natural compounds as the source of novel antiinfective drugs. Very promising materials, so far very little explored in anthelmintic drug discovery, are marine organisms. In this respect the organic compound, nafuredin is a very potent agent against nematodes due to specific inhibition of their essential metabolic pathways in nanomolar concentrations and a very low toxicity to mammalian cells.

Plant extracts are good starting point and the active principles that induced anthelmintic activities might be found in one or more classes of phytochemicals. The variations in activities of certain plant are due to the differences in the proportion of the active components responsible for the tested activity. Their proportions might be different in extracts obtained with different types of extraction solutions. The synergistic effect of several bioactive components is considered an important factor responsible for the variations in anthelmintic activity of extracts from different plants. Although a clear dose-dependent effect has been found for some plant extracts, the biochemical nature of the secondary compounds is also suspected to partly explain the variability of results found under in vitro and in vivo conditions. Only limited available data support the "indirect" hypothesis which relates the effects on the worms to an improved host immune response.

It is possible that extracts showing high efficacy in the range of 10–100 μg/ml in vitro contain a few compounds, which target different sites in helminths. Thus, plants with a high proportion of essential oils and flavonoids were the most effective against gastro-intestinal nematodes, however similar nematocidal effect of both components when tested individually, was observed in the range of milligrams. Two separate models of action could be attributed to the efficacy of essential oils in the treatment of protozoan and probably also helminth infections: combination of immunomodulatory and direct antiparasitic effects. Plants containing high amounts of polyphenols might also have applications for protecting proteins from degradation in the rumen, increasing the efficiency of microbial protein synthesis in the rumen and decreasing methane emission, for using as antioxidants, antibacterial as well as anthelmintic agents. Many studies concluded that the higher efficacy of a plant extract is usually achieved by increasing the dose or by repeated dosing for few days. Some of the active components (for example: condensed tannins) may have direct nematocidal and also anti-nutritional effects in livestock, such as reduced food intake and performance, therefore it is essential to validate both anthelmintic and side effects of isolated plant products. Flavonoids, in particular some of their components like thymol, have also exerted activity in vitro and in vivo against flatworms, whereas in nematodes their cuticle seems to prevent absorption necessary for further effects. Nevertheless, they can be absorbed via the oral route. Preliminary data obtained in flatworm models suggested their interaction with the worm-specific physiological processes. Other mode of anthelmintic activity was suggested for glycosides against filarial nematodes, whereas saponins probably affected permeability of the nematode cuticle as result of their surfactant properties. Higher plants-derived alkaloids so far examined, were shown to interfere with vital functions of tissue-dwelling nematode larvae at low concentrations. The special group of molecules isolated from several fruits tissues are cysteine proteases, which are able to selectively affect parasitic nematodes in vitro and showing high efficacy in vivo in monogastric animals. Therefore, through the knowledge and understanding gained from basic pharmacological research in in vitro and in vivo controlled studies, an array of bioactive molecules could be discovered for further clinical applications in human and veterinary parasitology.

References

Abdel-Ghaffar F, Semmler M, Al-Rasheid KAS, Strassen B, Fischer K, Aksu G, Klimpel S, Mehlhorn H (2011) The effects of different plant extracts on intestinal cestodes and on trematodes. Parasitol Res 108:979–984. doi:10.1007/s00436-010-2167-5

Abu-El-Ezz NM (2005) Effects of *Nigella sativa* and *Allium cepa* oils on *Trichinella spiralis* in experimentally infected rats. J Egypt Soc Parasitol 35:511–523

Ael-Banhawey M, Ashry MA, EL-Ansary AK, Aly SA (2007) Effect of *Curcuma longa* or parziquantel on *Schistosoma mansoni* infected mice liver—histological and histochemical study. Indian J Exp Biol 45:877–889

Akkari H, Darghouth MA, Ben Salem H (2008) Preliminary investigations of the anti-nematode activity of *Acacia cyanophylla* Lindl: excretion of gastrointestinal nematode eggs in lambs browsing *A. cyanophylla* with and without PEG or grazing native grass. Small Rumin Res 74:78–83. doi:10.1016/j.smallrumres.2007.03.012

Albuquerque AAC, Sorenson AL, Leal-Cardoso JH (1995) Effects of essential oil of *Croton zehntneri*, and of anethole and estragole on skeletal muscles. J Ethnopharmacol 49:41–49. doi:SSDI0378-8741(95)01301-S

Allam G (2009) Immunomodulatory effects of curcumin treatment on murine schistosomiasis mansoni. Immunobiology 214:712–727. doi:10.1016/j.imbio.2008.11.017

Al-Shaibani IRM, Phulan MS, Arijo A, Qureshi TA (2008) Ovicidal and larvicidal properties of *Adhatoda vasica* (L.) extracts against gastrointestinal nematodes of sheep in vitro. Pakistan Vet J 28: 79–83

Andlauer W, Furst P (2002) Nutraceuticals: a piece of history, present status and outlook. Food Res Int 35:171–176. doi:org/10.1016/S0963-9969(01)00179-X

Aremu AO, Fawole OA, Chukwujekwu JC, Light ME, Finnie JF, Van Staden J (2010) In vitro antimicrobial, anthelmintic and cyclooxygenase-inhibitory activities and phytochemical analysis of *Leucosidera sericea*. J Ethnopharmacol 131:22-27. doi: 10.1016/j.jep.2010.05.043

Aroche LU, Sánchez SDO, de Gives MP, Arellano LME, Hernandez LE, Cisneros VG, Ataide ADM, Velazquez HV (2008) In vitro nematicidal effects of medicinal plants from Sirra de Huautla, Biosphere Reserve, Morelos, Mexico against Haemonchus contortus infective larvae. J Helminth 82:25–31

Athanasiadou S, Kyriazakis I (2004) Plant secondary metabolites: antiparasitic effects and their role in ruminant production systems. Proc Nutr Soc 63:631–639. doi:10.1079/PNS2004396

Athanasiadou S, Kyriazakis I, Jackson F, Coop RL (2001) Direct anthelmintic effects of condensed tannins towards different gastrointestinal nematodes of sheep: in vitro and in vivo studies. Vet Parasitol 99:205–219. doi:org/10.1016/S0304-4017(01)00467-8

Athanasiadou S, Githiori J, Kyriazakis I (2007) Medical plants for helminth parasite control: facts and fiction. Animal 1:1392–1400. doi:org/10.1017/S1751731107000730

Ayers S, Zink DL, Mohn K, Powell JS, Brown CHM, Murphy T, Brand R, Pretorius S, Stevenson D, Thompson D, Singh SB (2008) Flavones from *Struthiola argentea* with anthelmintic activity in vitro. Phytochemistry 69:541–545. doi:10.1016/j.phytochem.2007.08.003

Azando EVB,Hounzangbe-Adote MS, Olounlade PA, Brunet S, Fabre N, Valentin A, Hoste H (2011) Involvement of tannins and flavonoids in the in vitro effects of *Newbouldia laevis* and *Zanthoxylum zanthoxyloides* extracts on the exsheathment of third-stage infective larvae of gastrointestinal nematodes. Vet Parasitol 180:292–297. doi:10.1016/j.vetpar.2011.03.010

Bakkali F, Averbeck S, Averbeck D, Zhiri A, Idaomar M (2005) Cytotoxicity and gene induction by some essential oils in the yeast *Saccharomyces cerevisiae*. Mutat Res 585:1–13

Bakkali F, Averbeck S, Averbeck D, Idaomar M (2008) Biological effects of essential oils—a review. Food Chem Toxicol 46:446–475. doi:10.1016/j.fct.2007.09.106

Baraldi R, Isacchi B, Predieri S, Marconi G, Vincieri FF, Bilia AR (2008) Distribution of artemisinin and bioactive flavonoids from *Artemisia annua* L. during plant growth. Biochem Syst Ecol 36:340–348. doi:10.1016/j.bse.2007.11.002

Barnes S (1998) Evolution of the health benefits of soy isoflavones. Proc Soc Exp Biol Med 217:386–392. doi:10.3181/00379727-217-44249

Barrau E, Fabre N, Fouraste I, Hoste H (2005) Effect of bioactive compounds from Sainfoin (*Onobrychis viciifolia* Scop.) on the in vitro larval migration of *Haemonchus contortus*: role of tannins and flavonol glycosides. Parasitology 131:531–538. doi:org/10.1017/S0031182005008024

Behnke JM, Buttle DJ, Stepek G, Lowe A, Duce IR (2008) Developing new anthelmintics from plant cysteine proteinases (review). Parasit Vectors 1:29. doi:10.1186/1756-3305r-r1-29

Bernes G, Waller PJ, Christensson D (2000) The effect of birdsfoot trefoil (*Lotus corniculatus*) and white clover (*Trifolium repens*) in mixed pasture swards on incoming and established nematode infections in young lambs. Acta Vet Scand 41:351–361

Bodiwala HS, Singh G, Singh R, Dey CS, Sharma SS, Bhutani KK, Singh IP (2007) Antileishmanial amides and lignans from *Piper cubeba* and *Piper retrofractum*. J Nat Med 61:418–421. doi:10.1007/s11418-007-0159-

Braguine CG, Bertanha CS, Goncalves UO, Magalhaes LG, Rodrigues V, Gimenez VMM, Groppo M, Silva MLAE, Cunha WR, Januario AH, Pauletti PM (2012) Schistosomicidal evaluation of flavonoids from two species of Styrax against *Schistosoma mansoni* adult worms. Pharmaceut Biol 50:925–929. doi:10.3109/13880209.2011.649857

Brooker S, Clements ACA, Bundy DAP (2006) Global epidemiology, ecology and control of soil-transmitted helminth infections. Global mapping of infectious diseases: Methods, Examples and emerging applications book series. Adv Parasitol 62:221–261. doi:10.1016/S0065-308X(05)62007-6

Brunet S, Hoste H (2006) Monomers of condensed tannins affect the larval exsheathment of parasitic nematodes of ruminants. J Agric Food Chem 54:7481–7487. doi:10.1021/jf0610007

Bryant C, Behm AC (1989) Biochemical adaptation in parasites. Chapman and Hall, London, pp 25–69

Buttle DJ, Behnke JM, Bartley Y, Elsheikha HM, Bartley DJ, Garnett MC, Donnan AA, Jackson F, Lowe A, Duce IR (2011) Oral dosing with papaya latex is an effective anthelmintic treatment for sheep infected with *Haemonchus contortus*. Parasit Vectors 4:36

Caixeta SC, Magalhães LG, Melo NI, Wakabayashi KAL, Aguiar GP, Aguiar DP, Mantovani ALL, Morais JA, Oliveira PF, Tavares DC, Groppo M, Rodrigues V, Cunha WR, Veneziani RCS, da Silva Filho AA, Crotti AEM (2011) Chemical composition and in vitro schistosomicidal activity of the essential oil of *Plectranthus neochilus* grown in Brazil Southeast. Chem Biodivers. doi:10.1002/cbdv.201100167

Cala AC, Chagas ACS, Oliveira MCS, Matos AP, Borges LMF, Sousa LAD, Souza FA, Oliveira GP (2012) In vitro anthelmintic effect of *Melia azedarach* L. and *Trichilia claussenii* C. against sheep gastrointestinal nematodes. Exp Parasitol 130:98–102. doi:10.1016/j.exppara.2011.12.011

Callahan HL, Crouch RK, James ER (1988) Helminth antioxidant enzymes: a protective mechanism against host oxidants? Parasitol Today 4:218–225

Camura-Vasconcelos ALF, Bevilaqua CML, Morais SM, Maciel MV, Costa CTC, Macedo ITF, Oliveira LMB, Braga RR, Silva RA, Vieira LS (2007) Anthelmintic activity of *Croton zehntneri* and *Lippia sidoides* essential oils. Vet Parasitol 148:288–294. doi:10.1016/j.vetpar.2007.06.012

Caner A, Döskaya M, Degirmenci A, Can H, Baykan S, Űner A, Basdemir G, Zeybek U, Gűrűz Y (2008) Comparison of the effects of *Artemisia vulgaris* and *Artemisia absinthium* growing in western Anatolia against trichinellosis (*Trichinella spiralis*) in rats. Exp Parasitol 119:173–179. doi:10.1016/j.exppara.2008.01.012

Cao CH, Liu Y, Lehmann M (2007) Fork head controls the timing and tissue selectivity of steroid-induced developmental cell death. JCB 176:843–852. doi:10.1083/jcb.200611155

Capon RJ, Vuong D, Lacey E, Gill JH (2005) (-)-Echinobetaine A: Isolation structure elucidation, synthesis, and SAR studies on a new nematocide from a southern Australian marine sponge, *Echinodictyum* sp. J Nat Prod 68:179–182. doi:10.1021/np049687h

Carvalhoa CO, Chagas ACS, Cotinguiba F, Furlan M, Brito LG, Chaves FCM, Stephan MP, Bizzo HR, Amarante AFT (2012) The anthelmintic effect of plant extracts on *Haemonchus contortus* and *Strongyloides venezuelensis*. Vet Parasitol 183:260–268. doi:10.1016/j.vetpar.2011.07.051

Challam M, Roy B, Tandon V (2010) Effect of *Lysimachia ramosa* (Primulaceae) on helminth parasites. Motility, mortality and scanning electron microscopic observations on surface topography. Vet Parasitol 169:214–218. doi:10.1016/j.vetpar.2009.12.024

Chatterjee RK, Fatma N, Murthy PK, Sinha P, Kulshrestha DK, Dhawan BN (1992) Macrofilaricidal activity of the stembark of *Streblus asper* and its major active constituents. Drug Dev Res 26:67–78

Chen DJ, Fu LF, Shao PP, Wu FZ, Fan CZ, Shu H, Ren CS, Sheng XL (1980) Studies on antischistosomal activity of qinghaosu in experimental therapy. Zhong Hui Yi Xue Zha Zhi 80:422–428 (in Chinese)

Cioli D, Pica-Mattoccia L (2003) Praziquantel. Parasitol Res 90:S3–S9. doi:10.1007/s00436-002-0751-z

Clark L, Cools R, Robbins TW (2004) The neuropsychology of ventral prefrontal cortex: decision-making and reversal learning. Brain Cogn 55:41–53. doi:10.1016/S0278-2626(03)00284-7

Conder GA, Jen LW, Marbury KS, Johnson SS, Guimond PM, Thomas EM, Lee BL (1990) A novel anthelmintic model utilizing *Meriones unguiculatus*, infected with *Haemonchus contortus*. J Parasitol 76:168–170. doi:10.2307/3283008

Crews P, Hunter LM (1993) The search for antiparasitic agents from marine animals. In: Attaway DH, Zaborsky OR (eds) Marine biotechnology. Plenum Press, New York, USA, London, UK, pp 343–389

Das B, Tandon V, Saha N (2004) Effects of phytochemicals of *Flemingia vestita* (Fabaceae) on glucose 6-phosphate dehydrogenase and enzymes of gluconeogenesis in a cestode *Raillietina echinobothrida*. Comp Biochem Physiol 139:141–146. doi:org/10.1016/j.cca.2004.10.004

Das B, Tandon V, Saha N (2006) Effect of isoflavone from *Flemingia vestita* (Fabaceae) on the Ca_2+ homeostasis in *Raillietina echinobothrida*, the cestode of domestic fowl. Parasitol Inter 55:17–21. doi:org/10.1016/j.parint.2005.08.002

Das B, Tandon V, Lyndem LL, Gray AI, Ferro VA (2009) Phytochemicals from *Flemingia vestita* (fabaceae) and *Stephania glabra (Menispermeaceae)* alter cGMP concentration in the cestode *Raillietina echinobothrida*. Comp Bioc Physiol 149:397–403. doi:org/10.1016/j.cbpc.2008.09.012

Dayan AD (2003) Albendazole, mebendazole and praziquantel. Review of non-clinical toxicity and pharmacokinetics. Acta Trop 86:141–159. doi:10.1016/S0001-706X(03)00031-7

de Amorin A, Borba HR, Carauta JP, Lopes D, Kaplan MA(1999) Anthelmintic activity of the latex of *Ficus species*. J Ethnopharmacol 64:255–258. PII S0378-8741/98/00139-4

De Gives PM, López Arellano ME, Hernández EL, Marcelino LA (2012) Plant extracts: a potential tool for controlling animal parasitic nematodes. In: The biosphere. ISBN: 978-953-51-0292-2. doi:10.5772/34/783 de Melo NI, Magalhaes LG, de Carvalho CE, Wakabayashi KAL, de P Aguiar G, Ramos RC, Mantovani ALL, Turatti ICC, Rodrigues V, Groppo M, Cunha WR (2011) Schistosomicidal activity of the essential oil of *Ageratum conyzoides* L. (Asteraceae) against adult *Schistosoma mansoni* worms. Molecules 16:762–73. doi:10.3390/molecules16010762

de Moraes J, Nascimento C, Lopes POMV, Nakano E, Yamaguchi LF, Kato MJ, Kawano T (2011) *Schistosoma mansoni*: in vitro schistosomicidal activity of piplartine. Exp Parasitol 127:357–364. doi:10.1016/j.exppara.2010.08.021

de Moraes J, Nascimento C, Yamaguchi LF, Kato MJ, Nakano E(2012) *Schistosoma mansoni* in vitro schistosomicidal activity and tegumental alterations induced by piplartine on schistosomula. Exp Parasitol (in press) doi:org/10.1016/j.exppara.2012.07.004

de Oliviera RN, Rehder VLG, Oliveira ASS, Montanari I Jr, de Carvalho JE, Gois de Ruiz ALT, Jeraldo VLS, Linhares AX, Allegretti SM (2012) *Schistosoma mansoni*: In vitro schistosomicidal activity of essential oil of *Baccharis trimera*. Exp Parasitol doi:10.1016/j.exppara.2012.06.005 (in press)

Diaz AMA, Acosta TJFJ, Castro SCA, Hoste H (2010) Tannins in tropical tree fodders fed to small ruminants: a friendly foe? Small Rumin Res 89:164–173. doi:10.1016/j.smallrumres.2009.12.040

Doligalska M, Jozwicka K, Kiersnowska M, Mroczek A, Paczkowski C, Janiszowska W (2011) Triterpenoid saponins affect the function of P-glycoprotein and reduce the survival of the free-living stages of *Heligmosomoides bakeri*. Vet Parasitol 179:144–151. doi:10.1016/j.vetpar.2011.01.053

Duthaler U, Smith TA, Keiser J (2010) In vivo and in vitro sensitivity of *Fasciola hepatica* to triclabendazole combined with artesunate, artemether, or OZ78. Antimicrob Agent Chemother 54:4596–4604. doi:10.1128/AAC.00828-10

Duthaler U, Huwyler J, Rinaldi L, Cringoli G, Keiser J (2012) Evaluation of the pharmacokinetic profile of artesunate, artemether and their metabolites in sheep naturally infected with *Fasciola hepatica*. Vet Parasitol 186:270–280. doi:10.1016/j.vetpar.2011.11.076

Egerton JR, Ostlind DA, Blair LS, Eary CH, Suhayda D, Cifeli S, Riek RF, Camphell WC (1979) Avermectins, new family of potent anthelmintic agents: efficacy of the B1α component. Antimicrob Agents Chemother 15:372–378

Eguale T, Tilahun G, Debella A, Feleke A, Makonnen E (2007a) In vitro and in vivo anthelmintic activity of crude extracts of *Coriandrum sativum* against *Haemonchus contortus*. J Ethnopharmacol 110:428–433. doi:10.1016/j.jep.2006.10.003

Eguale T, Tilahun G, Debella A, Feleke A, Makonnen E (2007b) *Haemonchus contortus*: in vitro and in vivo anthelmintic activity of aqueous and hydro-alcoholic extracts of *Hedera helix*. Exp Parasitol 116:340–345. doi:10.1016/j.exppara.2007.01.019

Elissondo MC, Albani CM, Gende L, Eguaras M, Denegri G (2008) Efficacy of thymol against Echinococcus granulosus protoscoleces. Parasitol Int 57:185–190. doi:10.1016/j.paraint.2007.12.005

EMEA-European Medicines Evaluation Agency (1996) Praziquantel summary report by CVMP. EMEA/MRL/141/96, Sept 1996. EMEA, London

EMEA-European Medicines Evaluation Agency (1997) Albendazole summary report by CVMP. EMEA/MKL/247/97-Final, EMEA, London

EMEA-European Medicines Evaluation Agency (1999) Mebendazole summary report by CVMP. EMEA/MKL/625/99-Final. July 1999. EMEA, London

Fandohan P, Gnonlonfin B, Laleye A, Gbenou JD, Darbouxc R, Moudachirou M (2008) Toxicity and gastric tolerance of essential oils from *Cymbopogon citratus*, *Ocimum gratissimum* and *Ocimum basilicum* in Wistar rats. Food Chem Toxicol 46:2493–2497. doi:10.1016/j.fct.2008.04.006

Ferreira JFS, Luthria DL, Sasaki T, Heyerick A (2010) Flavonoids from *Artemisia annua* L. as antioxidants and their potential synergism with artemisinin against malaria and cancer. Molecules 15:3135–3170. doi:10.3390/molecules15053135

Fester K (2010) Plant alkaloids. Published online. doi:10.1002/9780470015902.a0001914

Fioravanti CF, Walker DJ, Sandhu PS (1998) Metabolic transition in the development of *Hymenolepis diminuta* (Cestoda). Parasitol Res 84:777–782. doi:10.1007/s004360050487

Foster JG, Clapham WM, Belesky DP, Labreveux M, Hall MH, Sanderson MA (2006) Influence of cultivation site on sesquiterpene lactone composition of forage chicory (*Cichorium intybus* L.). J Agric Food Chem 54:1772–1778. doi:10.1021/jf052546g

Foster JG, Cassida KA, Turner KE (2011) In vitro analysis of the anthelmintic activity of forage chicory (*Cichorium intybus* L.) sesquiterpene lactones against a predominantly *Haemonchus contortus* egg population. Vet Parasitol 180:298–306. doi:10.1016/j.vetpar.2011.03.013

Francis G, Kerem Z, Makkar HPS, Becker K (2002) The biological action of saponins in animal systems: a review. Br J Nutr 88:587–605. doi:10.1079/BJN2002725

Frayha GJ, Smyth JD, Gobert JG, Savel J (1997) The mechanisms of action of antiprotozoal and anthelmintic drugs in man. Gen. Pharmacol 28:273–299. PII:S0306–S3623 (96)00149-8

Gabino JAF, de Gives MMP, Sanchez SME, Hernandez LE, Velazquez HVM, Cisneros VG (2010) Anthelmintic effects of Prosopsis laevigata n-hexanic extract against Haemonchus contortus in artificially infected gerbils (Meriones unguiculatus). J Helminth 84:71–75

Gaur RL, Sahoo MK, Dixit S, Fatma N, Rastogi S, Kulshreshtha DK, Chatterjee RK, Murthy PK (2008) Antifilarial activity of *Caesalpinia bonducella* against experimental filarial infections. Indian J Med Res 128:65–70

Geary TG, Sangster NC, Thompson DP (1999) Frontiers in anthelmintic pharmacology. Vet Parasitol 84:275–295. doi:org/10.1016/S0304-4017(99)00042-4

Geary TG, Woo K, McCarthy JS, Mackenzie CD, Horton J, Prichard RK, de Silva N, Olliaro PL, Lazdins-Helds JK, Engels DA, Bundy DA (2010) Unresolved issues in anthelmintic pharmacology for helminthiases of humans. Int J Parasitol 40:1–13. doi:10.1016/j.ijpara.2009.11.001

Ghosh NK, Babu SP, Sukul NC, Ito A (1996) Cestocidal activity of *Accacia auriculiformis*. J Helminthol 70:171–172. doi:10.1017/S0022149X00015340

Githiori JB, Athanasiadou S, Thamsborg SM (2006) Use of plants in novel approaches for control of gastrointestinal helminths in livestock with emphasis on small ruminants. Vet Parasitol 139:308–320. doi:10.1016/j.vetpar.2006.04.021

Gupta J, Misra S, Mishra SK, Srivastava S, Srivastava MN, Lakshmi V, Misra-Bhattacharya S (2012) Antifilarial activity of marine sponge *Haliclona oculata* against experimental *Brugia malayi* infection. Exp Parasitol 130:449–455. doi:10.1016/j.exppara.2012.01.009

Hale LP (2004) Proteolytic activity and immunogenicity of oral bromelain within the gastrointestinal tract of mice. Int Immunopharmacol 4:255–264. doi:10.1016/j.intimp.2003.12.010

Hämäläinen M, Nieminen R, Vuorela P, Heinonen M, Moilanen E (2007) Anti-inflammatory effects of flavonoids: Genistein, kaempferol, quercetin, and daidzein inhibit STAT-1 and NF—$_\kappa$B activations, wherwas flavone isorhamnetin, naringenin, and Pelargonidin inhibit only NF—$_\kappa$B activation along with their inhibitory effect on iNOS expression and NO production in activated macrophages. Mediat Inflamm. doi:10.1155/2007/45673

Hansson A, Veliz G, Naquira C, Amren M, Arroyo M, Arevalo G (1986) Preclinical and clinical studies with latex from *Ficus glabrata* HBK, a traditional intestinal anthelminthic in the Amazonian area. J Ethnopharamacol 17:105–138

Haq A, Abdullatif M, Lobo PI, Khabar KSA, Sheth KV, Al-Sedairy ST (1995) *Nigella sativa*: effect on human lymphocytes and polymorphnuclear leukocyte phagocytic activity. Immunopharmacology 30:147–155. SSDI 0162-3109(95)00016-X

Harder A, von Samson-Himmelstjerna G (2002) Cyclooctadepsipeptides—a new class of anthelmintically active compounds. Parasitol Res 88:481–488

Havsteen BH (2002) The biochemistry and medical significance of the flavonoids. Pharmacol Ther 96: 67-202. doi:PII:S0163-7258(02)00298-X

Hernández-Villegas MM, Borges-Argáez R, Rodríguez-Vivas RI, Torres-Acosta JFJ, Méndez-Gonzáles M, Cáceres-Farfán M (2011) Ovicidal and larvicidal activity of the crude extracts from *Phytolacca icosandra* against *Haemonchus contortus*. Vet Parasitol 179:100–106. doi:10.1016/j.vetpar.2011.02.019

Hernández-Villegas MM, Borges-Argáez R, Rodríguez-Vivas RI, Torres-Acosta JFJ, Méndez-Gonzáles M, Cáceres-Farfán M (2012) In vivo anthelmintic activity of *Phytolacca icosandra* against *Haemonchus contortus* in goats. Vet Parasitol doi:org/10.1016/j.vetpar.2012.04.017 (in press)

Holden-Dye L, Walker RJ (2007) Anthelmintic drugs. In: Maricq V, McIntire L (eds) WormBook. The *C. elegans* research community. doi:10.1895/wormbook.1.143.1

Hördegen P, Hertzberg H, Heilmann J, Langhans W, Maurer V (2003) The anthelmintic efficacy of five plant products against gastrointestinal trichostrongylids in artificially infected lambs. Vet Parasitol 117:51–60. doi:10.1016/j.vetpar.2003.07.027

Hoste H, Torres-Acosta JFJ (2011) Non chemical control of helminths in ruminants: adapting solutions for changing worms in a changing world. Vet Parasitol 180:144–154. doi:10.1016/j.vetpar.2011.05.085

Hoste H, Torres-Acosta JF, Paolini V, Aguilar-Caballero A, Etter E, Lefrileux Y, Chartier C, Broqua C (2005) Interactions between nutrition and gastrointestinal infections with parasitic nematodes in goats. Small Ruminant Res 60:141–151. doi:10.1016/j.smallrumres.2005.06.008

Hoste H, Jackson F, Athanasiadou S, Thamsborg SM, Hoskin SO (2006) The effects of tannin-rich plants on parasitic nematodes in ruminants. Trends Parasitol 22:253–261. doi:org/10.1016/j.pt.2006.04.004

Hoste H, Martinez-Ortiz-De-Montellano C, Manolaraki F, Brunet S, Ojeda-Robertos N, Fourquaux I, Torres-Acosta JFJ, Sandoval-Castro CA (2012) Direct and indirect effects of bioactive tannin-rich tropical and temperate legumes against nematode infections. Vet Parasitol 186:18–27. doi:10.1016/j.vetpar.2011.11.042

Hotez PJ, Brindley PJ, Bethony JM, King CH, Pearce EJ, Jacobson J (2008) Helminth infections: the great neglected tropical diseases. J Clin Investig 118:1311–1321. doi:10.1172/JCI34261

Hu M, Konoki K, Tachibana K (1996) Cholesterol-independent membrane disruption caused by triterpenoid saponins. Biochim Biophys Acta 1299:252–258

Iqbal Z, Khahid-Nadeem Q, Kham MN, Akthar MSS, Waraich FN (2001) In vitro anthelmintic activity of *Allium sativum, Zingiber officinale* and *Ficus religiosus*. Int J Agric Biol 3:454–457

Iqbal Z, Lateef M, Ashraf M, Jabbar A (2004) Anthelmintic activity of *Artemisia brevifolia* in sheep. J Ethnopharmacol 93:265–268. doi:10.1016/j.jep.2004.03.046

Itakura Y, Ichikawa M, Mori Y, Okino R, Udayama M, Morita T (2001) How to distinguish garlic from the other *Allium* vegetables. J Nutr 131:963S–967S

Jabbar A, Zaman MA, Iqbal Z, Yassen M, Shamim A (2007) Anthelmintic activity of *Chemopodium album* (L) and *Caesalpinia crista* (L) against trichostrongylid nematodes of sheep. J Ethnopharmacol 114: 86–91

Janse CJ, Waters AP, Kos J, Lugt CB (1994) Comparison of in vivo and in vitro antimalarial activity of artemisinin, dihydroartemisinin and sodium artesunate in the *Plasmodium berghei* rodent model. Int J Parasitol 24:589–594. doi:org.proxy.library.ucsb.edu:2048/10.1016/0020-7519(94)90150-3

Kahiya C, Mukaratirwa S, Thamsborg SM (2003) Effects of *Acacia nilotica* and *Acacia karoo* diet on *Haemonchus contortus* infection in goats. Vet Parasitol 115:265–274. doi:10.1016/S0304-4017(03)00213-9

Kamaraj C, Rahuman AA (2011) Efficacy of anthelmintic properties of medicinal plant extracts against *Haemonchus contortus*. Res Vet Sci 91:400–404. doi:10.1016/j.rvsc.2010.09.018

Kamaraj C, Rahuman AA, Bagavan A, Mohamed MJ, Elango G, Rajakumar G, Zahir AA, Santhoshkumar T, Marimuthu S (2010a) Ovicidal and larvicidal activity of crude extracts of *Melia azedarach* against *Haemonchus contortus* (Strongylida). Parasitol Res 106:1071–1077. doi:10.1007/s00436-010-1750-0

Kamaraj C, Rahuman AA, Bagavan A, Elango G, Rajakumar G, Zahir AA, Marimuthu S, Santhoshkumar T, Jayaseelan C (2010b) Evaluation of medicinal plant extracts against blood-sucking parasites. Parasitol Res 106:1403–1412. doi:10.1007/s00436-010-1816-z

Kaplan RM (2004) Drug resistance in nematodes of veterinary importance: a status report. Trends Parasitol 20:477–481. doi:org/10.1016/j.pt.2004.08.001

Kar P, Tandon V, Saha N (2002) Anthelmintic efficacy of *Flemingia vestita*: genistein-induced effect on the activity of nitric oxide syntase and nitric oxide in the trematode parasite, *Fasciolopsis buski*. Parasitol Int 51:249–257. PII:S1383-5769(02)00032-6

Kar PK, Tandon V, Saha N (2004) Anthelmintic efficacy of genistein, the active principle of *Flemingia vestita* (Fabaceae): alterations in the free amino acid pool and ammonia levels in the fluke *Fasciolipsis bruski*. Parasitol Int 53:287–291. doi:10.1016/j.parint.2004.04.001

Kasinathan RS, Morgan WM, Greenberg RM (2010) *Schistosoma mansoni* express higher levels of multidrug resistance-associated protein 1 (SmMRP1) in juvenile worms and in response to praziquantel. Mol Biochem Parasitol 173:25–31. doi:10.1016/j.molbiopara.2010.05.003

Katiki LM, Ferreira JFS, Zajac AM, Masler C, Lindsay DS, Chagas ACS, Amarante AFT (2011a) *Caenorhabditis elegans* as a model to screen plant extracts and compounds as natural anthelmintics for veterinary use. Vet Parasitol 182:264–268. doi:10.1016/j.vetpar.2011.05.020

Katiki LM, Chagasb ACS, Bizzoc HR, Ferreirad JFS, Amarantee AFT (2011b) Anthelmintic activity of *Cymbopogon martinii*, *Cymbopogon schoenanthus* and *Mentha piperita* essential oils evaluated in four different in vitro tests. Vet Parasitol 183:103–108. doi:10.1016/j.vetpar.2011.07.001

Katiki LM, Chagasb ACS, Takahirac RK, Juliani HR, Ferreirac JFS, Amarante AFT (2012) Evaluation of *Cymbopogon schoenanthus* essential oil in lambs experimentally infected with *Haemonchus contortus*. Vet Parasitol 186:312–318. doi:10.1016/j.vetpar.2011.12.003

Keiser J, Morson G (2008) *Fasciola hepatica*: tegumental alterations in adult flukes following in vitro and in vivo administration of artesunate and artemether. Exp Parasit 118:228–237. doi:10.1016/j.exppara.2007.08.007

Keiser J, Utzinger J (2005) Emerging foodborne trematodiasis. Emer Inf Dis 11:1507–1514. doi:10.1016/j.vetpar.2010.09.011

Keiser J, Utzinger J (2007) Artemisinins and synthetic trioxolanes in the treatment of helminth infections (review). Curr Opin Infect Dis 20:605–612. doi:10.1097/QCO.0b013e3282f19ec4

Keiser J, Utzinger J (2010) The drugs we have and the drugs we need against major helminth infections—chapter 8. Adv Parasitol 73:197–230. http://dx.doi.org/10.1016/S0065-308X(10)73008-6

Keiser J, Vargas M (2010) Effect of artemether, artesunate, OZ78, praziquantel, and tribendimidine alone or in combination chemotherapy on the tegument of *Clonorchis sinensis*. Parasitol Int 59:472–476. doi:10.1016/j.parint2010.04.003

Keiser J, Xiao SH, Tanner M, Utzinger J (2006a) Artesunate and artemether are effective fasciolicides in the rat model and in vitro. J Antimicrob Chemother 57:1139–1145. doi:10.1093/jac/dkl125

Keiser J, Xiao SH, Xue J, Chang ZS, Odermatt P, Tesana S, Tanner M, Utzinger J (2006b) Effect of artesunate and artemether against *Clonorchis sinensis* and *Opisthorchis viverrini* in rodent models. Int J Antimicrob Agent 28:370–373. doi:10.1016/j.ijantimicag.2006.08.004

Keiser J, Rinaldi L, Veneziano V, Mezzino L, Tanner M, Utzinger J, Cringoli G (2008) Efficacy and safety of artemether against a natural *Fasciola hepatica* infection in sheep. Parasitol Res 103:517–522. doi:10.1007/s00436-008-0998-0

Keiser J, Veneziano V, Rinaldi L, Mezzino L, Duthaler U, Cringoli G (2010) Anthelmintic activity of artesunate against *Fasciola hepatica* in naturally infected sheep. Res Vet Sci 88:107–110. doi:10.1016/j.rvsc.2009.05.007

Keiser J, Sayed H, El-Ghanam M, Sabry H, Anani S, El-Wakeel A, Hatz Ch, Utzinger J, Seif el Din S, El-Maadawy W, Botros S (2011) Efficacy and safety of artemether in the treatment of chronic Facioliasis in Egypt: exploratory phase-2 trials. PloS Negl Trop Dis 5:e1285. doi:10.1371/journal.pntd.0001285

Kennedy MW, Foley M, Kuo YM, Kusel JR, Garland PB (1987) Biophysical properties of the surface lipid of parasitic nematodes. Mol Biochem Parasitol 22:233–240. doi:10.1016/0166-6851(87)90054-5

Kerboeuf D, Guegnard F (2011) Anthelmintics are substrates and activators of nematode P Glycoprotein. Antimicrob Agent Chemother 55:2224–2232. doi:10.1128/AAC.01477-10

Kerboeuf D, Blackhall W, Kaminsky R, von Samson-Himmelstjerna G (2003) P-glycoprotein in helminths: function and perspectives for anthelmintic treatment and reversal of resistance. Int J Antimicrob Agents 22:332–346. doi:10.1016/S0924-8579(03)00221-8

Kerboeuf D, Riou M, Neveu C, Issouf M (2010) Membrane drug transport in helminths. Anti-infect. Agent Med Chem 9:113–129. doi:10.1139/O09-126

Ketzis JK, Taylor A, Bowman DD, Brown DL, Warnick LD, Erb HN (2002) *Chenopodium ambrosioides* and its essential oil as treatments for *Haemonchus contortus* and mixed adult-nematode infections in goats. Small Rumin Res 44:193–200. PII:S0921-4488/(02)00047-0

Kim TI, Yoo WG, Li S, Hong ST, Keiser J, Hong SJ (2009) Efficacy of artesunate and artemether against *Clonorchis sinensis* in rabbits. Parasitol Res 106:153–156. doi:10.1007/s00436-009-1641-4

Kita K, Shiomi K, Omura S (2007) Advances in drug discovery and biochemical studies (review). Trends Parasitol 23:223–229. doi:10.1016/j.pt.2007.03.005

Klein CB, King AA (2007) Genistein genotoxicity: Critical considerations of in vitro exposure dose. Toxicol Appl Pharmacol 224:1–11. doi:10.1016/j.taap.2007.06.022

Klimpel S, Abdel-Ghaffar FA, Al-Rasheid KAS, Aksu G, Fischer K, Strassen B, Melhorn H (2011) The effects of different plant extracts on nematodes. Parasitol Res 108:1047–1054. doi:10.1007/s00436-010-2168-4

Kohn AB, Roberts-Misterly JM, Anderson PAV, Khan N, Greenberg RM (2003) Specific sites in the beta interaction domain of a schistosome Ca^{2+} channel β-subunit are key to its role in sensitivity to the anti-schistosomal drug praziquantel. Parasitology 127:349–356. doi:10.1017/S003118200300386X

Königová A, Hrčkova G, Velebný S, Čorba J, Várady M (2008) Experimental infection of *Haemonchus contortus* strains resistant and susceptible to benzimidazoles and the effect on mast cells distribution in the stomach of Mongolian gerbils (*Meriones unguiculatus*). Parasitol Res 102:587–595. doi:10.1007/s00436-007-0792-4

Lacey E (1990) Mode of action of benzimidazoles. Parasitol Today 6:112–115

Lakshmi V, Kumar R, Gupta P, Varshney V, Srivastava MN, Dikshit M, Murthy PK, Misra-Bhattacharya S (2004a) The antifilarial activity of a marine red alga, *Botryocladia leptopoda*, against experimental infections with animal and human filariae. Parasitol Res 93:468–474. doi:10.1007/s00436-004-1159-8

Lakshmi V, Saxena A, Pandey K, Preeti Bajpai, Misra-Bhattacharya S (2004b) Antifilarial activity of *Zoanthus* species (Phylum Coelenterata, Class Anthzoa) against human lymphatic filaria, *Brugia malayi*. Parasitol Res 93:268–273. doi: 10.1007/s00436-004-1124-6

Lakshmi V, Joseph SK, Srivastava S, Verma SK, Sahoo MK, Dube V, Mishra SK, Murthy PK (2010) Antifilarial activity in vitro and in vivo of some flavonoids tested against *Brugia malayi*. Acta Trop 116:127–133. doi:10.1016/j.actatropica.2010.06.006

Lespine A, Ménez C, Bourguinat C, Prichard RK (2012) P-glycoproteins and other multidrug resistance transporters in the pharmacology of anthelmintics: prospects for reversing transport-dependent anthelmintic resistance (invited review). Int J Parasitol Drugs Drug Resist 2:58–75. doi:10.1016/j.ijpddr.2011.10.001

Li QG, Peggins JO, Fleckenstein LL, Masonic K, Heiffer MH, Brewer TG (1998) The pharmacokinetics and bioavailability of dihydroartemisinin, arteether, artemether, artesunic acid and artelinic acid in rats. J Pharm Pharmacol 50:173–182

Li YS, Chen HG, He HB, Hou XY, Ellis M, McManus DP (2005) A double-blind field trial on the effects of artemether on *Schistosoma japonicum* infection in a highly endemic focus in southern China. Acta Trop 96:184–190. doi:10.1016/j.actatropica.2005.07.013

Liu L, Song G, Hu Y (2007) GC–MS analysis of the essential oils of *Piper nigrum* L. and *Piper longum* L. Chromatographia 66:785–790. doi:10.1365/s10337-007-0408-2

Lotfy WM (2009) Human schistosomiasis in Egypt: historical rewiew, assessment of the current picture and prediction of the future trends. J Med Res Inst 30:1–7

Loukas A, Hotez PJ (2005) Chemotherapy of helminth infections. In: Brunton LL, Lazo JS, Parker KL (eds) Goodman & Gilman's the pharmacological basis of therapeutics, 11th edn. McGraw-Hill Companies, USA, pp 1073–1093

Luz PP, Magalhães LG, Pereira AK, Cunha WR, Rodrigues V, Marcio L. Andrade E Silva (2012) Curcumin-loaded into PLGA nanoparticles. Preparation and in vitro schistosomicidal activity. Parasitol Res 110:593–598 doi:10.1007/s00436-011-2527-9

Macedo ITF, Bevilaqua CML, de Oliveira LMB, Camurca-Vasconcelos ALF, Vieira SL, Oliveira FR, Queiroz-Junior EM, Tomé RA, Nascimento NRF (2009) Atividade ovicida e larvicida in vitro do óleo essencial de Eucalyptus globulus sobre Haemonchus contortus. Rev Bras Parasitol Vet 18:62–66. doi:org/10.4322/rbpv.01803011

Macedo ITF, Bevilaqua CML, de Oliveira LMB,Camurca-Vasconcelos ALF, Vieira SL, Oliveira FR, Queiroz-Junior EM, Tomé RA, Nascimento NRF (2010) Anthelmintic effect of Eucalyptus staigeriana essential oil against goat gastrointestinal nematodes. Vet Parasitol 173:93–98. doi:10.1016/j.vetpar.2010.06.004

Macedo IFT, Bevilaqua CML, de Oliveira LMB, Camurca-Vasconcelos ALF, Viera SL, Amóra SSA (2011) Evaluation of Eucalyptus citriodora essential oil on goat gastrointestinal nematodes. Rev Bras Parasitol Vet 20:223–227

Magalhaes LG, Lizandra G, de Souza JM, Wakabayashi KAL, Laurentiz RD, Vinholis AHC, Rezende KCS, Simaro GV, Bastos JK, Rodrigues V, Esperandim VR, Ferreira DS, Crotti AEM, Cunha WR, Silva MLAE (2012) In vitro efficacy of the essential oil of Piper cubeba L. (Piperaceae) against Schistosoma mansoni. Parasitol Res 110:1747–1754. doi:10.1007/s00436-011-2695-7

Magalhães LG, Machado CB, Morais ER, Bueno de Carvalho Moreira E, Sossai Soares C, Henrique da Silva S, Da Silva Filho AA, Rodrigues V (2009) In vitro schistosomicidal activity of curcumin against Schistosoma mansoni adult worms. Parasitol Res 104:1197–1201. doi:10.1007/s00436-008-1311-y

Magalhães LG, Kapadia GJ, Tonuci LRS, Caixeta SC, Parreira NA, RodriguesV, Filho AAS (2010) In vitro schistosomicidal effects of some phloroglucinol derivatives from Dryopteris species against Schistosoma mansoni adult worms. Parasitol Res 106:395–401. doi:10.1007/s00436-009-1674-8

Maheshwari RK, Singh AK, Gaddipati J, Srimal RC (2006) Multiple biological activities of curcumin: a short review. Life Sci 78:2081–2087. doi:10.1016/j.lfs.2005.12.007

Mahmoud MR, El-Abhar HS, Saleh S (2002) The effect of Nigella sativa oil against the liver damage induced by Schistosoma mansoni infection in mice. J Ethnopharmacol 79:1–11. PII: S0378-8741(01)00310-5

Martin RJ (1997) Modes of action of anthelmintic drugs. Vet J 154:11–34. doi:1(190-0233/t)7/04(1011-24/S12.00/0

Makkar HPS, Francis G, Becker K (2007) Bioactivity of phytochernicals in some lesser-known plants and their effects and potential applications in livestock and aquaculture production systems. Animal 1:1371–139. doi:10.1017/S1751731107000298

Martin R, Pennington AJ (1988) Effect of dihydroavermectin B1α on chloride-single-channel currents in Ascaris muscle. Pestic Sci 24: 90–91

Martin RJ, Robertson AP (2007) Mode of action of levamisole and pyrantel, anthelmintic resistance, E153 and Q57. Parasitology 134:1093–1104. doi:10.1017/S0031182007000029

Max RA (2010) Effect of repeated wattle tannin drenches on worm burdens, faecal egg counts and egg hatchability during naturally acquired nematode infections in sheep and goats. Vet Parasitol 169:138–143. doi:10.1016/j.vetpar.2009.12.022

Max RA, Wakelin D, Dawson J, Kimambo AE, Kassuku AA, Mtenga LA, Buttery PJ (2005) Effect of quebracho tannin on faecal egg counts, worm burdens and performance of temperate sheep with experimental nematode infections. J Agric Sci 143:519–527

Mayer AMS, Hamann MT (2005) Marine pharmacology in 2001–2002: marine compounds with anthelmintic, antibacterial, anticoagulant, antidiabetic, antifungal, anti-inflammatory, anti-malarial, antiplatelet, antiprotozoal, antituberculosis, and antiviral acties; affectin the cardiovascular, immune and nervous system and other miscellaneous mechanisms of action. Comp Biochem Physiol 140:265–286. doi:10.1016/j.cca.2005.04.004

McKellar AQ, Jackson F (2004) Veterinary anthelmintics: old and new. Trends Parasitol 20:456–461. doi:10.1016/j.pt.2004.08.002

Mehlhorn H, Al-Quraishy S, Al-Rasheid KAS, Jatzlau A, Abdel-Ghaffar F (2011a) Addition of a combination of anion (*Allium cepa*) and coconut (*Cocos nucifera*) to food of sheep stops gastrointestinal helmintic infection. Parasitol Res 108:1041–1046. doi:10.1007/s00436-010-2169-3

Mehlhorn H, Aksu G, Fischer K, Strassen B, Ghaffar FA, Al-Rasheid KAS, Klimpel S (2011b) The efficacy of extracts from plants- especially from coconut and onion—gainst tapeworms, trematodes, and nematodes. Nature helps-how plants and other organisms contribute to solve health problems. Book Ser Parasitol Res 1:109–139. doi:10.1007/978-3-642-19382-8_5

Meshnick SR (2002) Artemisinin: mechanisms of action, resistance and toxicity. Int J Parasitol 32:1655–1660. doi:10.1016/S0020-7519(02)00194-7

Middleton E, Kandaswami CH, Theoharides TC (2000) The effects of plant flavonoids on mammalian cells: Implications for inflammation, heart disease, and cancer. Pharmacol Rev 52: 673–751

Milgate J, Roberts DCK (1995) The nutritional and biological significance of saponins. Nutr Res 15:1223–1249. doi:org/10.1016/0271-5317(95)00081-S

Min BR, Hart SP (2003) Tannins for suppression of internal parasites. J Anim Sci 81:E102–E109

Min BR, Barry TN, Attwood GT, McNabb WC (2003) The effect of condensed tannins on the nutrition and health of ruminants fed fresh temperate forages: a review. Anim Feed Sci Technol 106:3–19. doi:10.1016/S0377-8401(03)00041-5

Misra N, Sharma M, Raj K, Dangi A, Srivastava S, Mishra-Bhattacharya S (2007) Chemical constituents and antifilarial activity of *Lantana camara* against human lymphatic filariid *Brugia malayi* and rodent filariid *Acanthocheilonema viteae* maintained in rodent models. Parasitol Res 100:439–448. doi:10.1007/s00436-006-0312-y

Misra S, Verma M, Mishra SK, Srivastava S, Lakshmi V, Misra-Bhattacharya S (2011) Gedunin and photogedunin of *Xylocarpus granatum* possess antifilarial activity against human lymphatic filarial parasite *Brugia malayi* in experimental rodent host. Parasitol Res 109:1351–1360. doi:10.1007/s00436-011-2380-x

Miyadera H, Shiomi K, Ui H, Yamaguchi Y, Masuma R, Tomoda H, Miyoshi H, Osanai A, Kita K, Ōmura S (2003) Atpenins, potent and specific inhibitors of mitochondrial complex II (succinateubiquinone oxidoreductase). PNAS 100:473–477. doi:10.1073/pnas.0237315100

Moazeni M, Saharkhiz MJ, Hosseini AA (2012) In vitro lethal effect of ajowan (*Trachyspermum ammi* L.) essential oil on hydatid cyst protoscoleces. Vet Parasitol 187:203–208. doi:10.1016/j.vetpar.2011.12.025

Molan AL, Waghorn GC, Min BM, McNabb WC (2000a) The effect of condensed tannins from seven herbages on *Trichostrongylus colubriformis* larval migration in vitro. Folia Parasitol 47:39–44. doi:10.1136/vr.150.3.65

Molan AL, Hoskin SO, Barry TN, McNabb WC (2000b) Effect of condensed tannins extracted from four forages on the viability of the larvae of deer lungworms and gastrointestinal nematodes. Vet Rec 147:44–48. doi:0.1136/vr.147.2.44

Molan AL, Alexander RA, Brookes IM, McNabb WC (2000c) Effect of an extract from sulla (*Hedysarum coronarium*) containing condensed tannins on the migration of three sheep gastrointestinal nematodes in vitro. Proc N Z Soc Anim Prod 60:21–25

Molan AL, Waghorn GC, McNabb WC (2002) The impact of condensed tannins on egg hatching and larval development of *Trichostrongylus colubriformis* in vitro. Vet Rec 150:65–69

Molan AL, Meagher LP, Spencer PA, Sivakumaran S (2003a) Effect of flavan-3-ols on in vitro egg hatching, larval development and viability of infective larvae of *Trichostrongylus colubriformis*. Int J Parasitol 33:1691–1698. doi:10.1016/S0020-7519(03)00207-8

Molan AL, Duncan AJ, Barry TN, McNabbWC (2003b) Effects of condensed tannins and crude sesquiterpene lactones extracted from chicory on the motility of larvae of deer lungworm and gastrointestinal nematodes. Parasitol Int 52:209–218. doi:10.1016/S1383-5769(03)00011-4

Mostafa OMS, Soliman MI (2010) Ultrastructure alterations of adult male of *Schistosoma mansoni* harbored in albino mice treated with Sidr honey and/or *Nigella sativa* oil. J. King Saud University (Sci) 22:111–121

Mueller-Harvey I (2006) Unravelling the conundrum of tannins in animal nutrition and health. J Sci Food Agric 86:2010–2037. doi:10.1002/jsfa.2577

Muraleedharan KM, Avery MA (2009) Progress in the development of peroxide- based anti-parasitic agents (review). Drug Disc Today 14:15–16. doi:10.1016/j.drudis.2009.05.008

Murrell KD, Pozio E (2011) Worldwide occurrence and impact of human trichinellosis 1986–2009. Emerg Inf Dis 17:2194–2202. doi:10.3201/eid1712.110896

Muthusamya VS, Ananda S, Sangeethaa KN, Sujathaa S, Arunb Balakrishnan, Lakshmi BS (2008) Tannins present in *Cichorium intybus* enhance glucose uptake and inhibit adipogenesis in 3T3-L1 adipocytes through PTP1B inhibition. Chem-Biol Interact 174:69–78. doi:10.1016/j.cbi.2008.04.016

Nagulesvaran A, Spicher M, Voniaufen N, Ortega-Mora LM, Torgerson P, Gottstein B, Hemphill A (2006) In vitro metacestodicidal activities of genistein and other isoflavones against *Echinococcus multilocularis* and *Echinococcus granulosus*. Antimicrob Agents Chemother 50:3770–3778. doi:10.1128/AAC.00578-06

Nandi B, Roy S, Bhattacharya S, Babu SPS (2004) Free radicals mediated membrane damage by the saponins acaciaside A and acaciaside B. Phytother Res 18:191–194. doi:10.1002/ptr.1387

Navickiene HMD, Alécio AC, Kato MJ, Bolzani VD, Young MC, Cavalheiro AJ, Furlan M (2000) Antifungal amides from *Piper hispidum* and *Piper tuberculatum*. Phytochemistry 55:621–626. doi:org/10.1016/S0031-9422(00)00226-0

Navickiene HMD, Bolzani VS, Kato MJ, Pereira AM, Bertoni BW, França SC, Furlan M (2003) Quantitative determination of anti-fungal and insecticide amides in adult plants, plantlets and callus from *Piper tuberculatum* by reversephase high-performance liquid chromatography. Phytochem Anal 14:281–284. doi:10.1002/pca.716

Nery PS, Nogueira FA, Martins ER, Duarte ER (2010) Effects of *Anacardium humile* leaf extracts on the development of gastrointestinal nematode larvae of sheep. Vet Parasitol 171:361–364. doi:10.1016/j.vetpar.2010.03.043

Nontprasert A, Pukrittayakamee S, Dondorp AM, Clemens R, Looareesuwan S, White NJ (2002) Neuropathologic toxicity of artemisinin derivates in a mouse model. Amer J Trop Med Hyg 67:423–429

Novobilský A, Mueller-Harvey I, Thamsborg SM (2011) Condensed tannins act against cattle nematodes. Vet Parasit 182:213–220. doi:10.1016/j.vetpar.2011.06.003

O'Neill JF, Johnston RC, Halferty L, Brennan GP, Keiser J, Fairweather I (2009) Adult triclabendazole-resistant *Fasciola hepatica*: morphological changes in the tegument and gut following in vivo treatment with artemether in the rat model. J Helminthol 83:151–163. doi:10.1017/S0022149X09344934

Oliveira LMB, Bevilaqua CML, Costa CTC, Macedo ITF, Barros RS, Rodrigues ACM, Camurca-Vasconcelos ALF, Morais SM, Lima YC, Vieira LS, Navarro AMC (2009) anthelmintic activity of *Cocos nucifera* L. against sheep gastrointestinal nematodes. Vet Parasitol 159:55–59. doi:10.1016/j.vetpar.2008.10.018

Olliaro PL, Haynes RK, Meunier B, Yuthavong Y (2001) Possible modes of action of the artemisinin-type compounds. Trends Parasitol 17 PII: S1471-4922(00)01838-X PII: S0020-7519(01)00297-1

Ōmura S (2002) Mode of action of avermectin. In Omura S (ed) Macrolide antibiotics. Chemistry, biology, and practice, 2nd edn. Academic Press, San Diego, pp 571–576

Ōmura S, Miyadera H, Ui H, Shiomi K, Yamaguchi Y, Masuma R, Nagamitsu T, Takano D, Sunazuka T, Harder A, Kölbl H, Namikoshi M, Miyoshi H, Sakamoto K, Kita K (2001) An anthelmintic compound, nafuredin, shows selective inhibition of complex I in helminth mitochondria. Proc Natl Acad Sci USA 98:60–62. doi:10.1073/pnas.011524698

Osbourn A (1996) Saponins and plant defence—a soap story. Trends Plant Sci 1:4–9. doi:10.1016/S13601385(96)80016-1

Osbourn A, Goss RJM, Field RA (2011) The saponins—polar isoprenoids with important and diverse biological activities. Nat Prod Rep 28:1261–1268. doi:10.1039/c1np00015b

Pal P, Tandon V (1998) Anthelmintic efficacy of *Flemingia vestita* (Leguminoceae): genistein-induced alterations in the activity of tegumental enzymes in the cestode, *Raillietina echinobothrida*. Parasitol Int 47:233–243. doi:org/10.1016/S1383-5769(98)00025-7

Parreira NA, Magalhães LG, Morais DR, Caixeta SC, de Sousa JPB, Bastos JK, Cunha WR, Silva MLA, Nanayakkara NPD, Rodrigues V, da Silva Filho AA (2010) Antiprotozoal, schistosomicidal, and antimicrobial activities of the essential oil from the leaves of *Baccharis dracunculifolia*. Chem Biodivers 7:993–100

Perkins S, Verschoyle RD, Hill K, Parveen I, Threadgill MD, Sharma RA, Williams ML, Steward WP, Gescher AJ (2002) Chemopreventive efficacy and pharmacokinetics of curcumin in the min/+ mouse, a model of familial adenomatous polyposis. Cancer Epidemiol Biomarkers Prev 11:535–540

Pessoa LM, Morais SM, Bevilaqua CML, Luciano JHS (2002) Anthelmintic activity of essential oil of *Ocimum gratissimum* Linn. and eugenol against *Haemonchus contortus*. Vet Parasitol 109:59–63. doi:PII:S0304-4017(02)00253-4

Pfarr KM, Qazi S, Fuhrman JA (2001) Nitric oxide synthase in filariae: demonstration of nitric oxide production by embryos in *Brugia malayi* and *Acanthocheilonema viteae*. Exp Parasitol 97:205–214. doi:10.1006/expr.2001

Pilatova M, Stupakova V, Varinska L, Sarissky M, Mirossay L, Mirossay A, Gal P, Kraus V, Dianiskova K, Mojzis J (2010) Effect of selected flavones on cancer and endothelial cells. Gen Physiol Biophys 29:134–143. doi:10.4149/gpb.2010.02.134

Poné JW, Tankoua OF, Yondo J, Komtangi MC, Mbida M, Bilong BCF (2011) The in vitro effects of aqueous and ethanolic extracts of the leaves of *Ageratum conyzoides* (Asteraceae) on three life cycle stages of the parasitic nematode *Heligmosomoides bakeri* (Nematoda: Heligmosomatidae). Vet Med Int 140293:5. doi:10.4061/2011/140293

Pozio E, La Rosa G, Morales MAG (2001) Epidemiology of human and animal trichinellosis in Italy since its discovery in 1887. Parasite 8:S106–S108

Prichard R, Ménez C, Lespine A (2012) Moxidectin and avermectins: consanguinity but not identity. Int J Parasitol Drugs Drug Resist 2:134–153. http://dx.doi.org/10.1016/j.jpddr.2012.04.001

Ramadan MF, Kroh LW, Morsel JT (2003) Radical scavenging activity of black cumin (*Nigella sativa* L.), coriander (*Coriandrum sativum* L.), and niger (*Guizotia abyssinica* Cass.) crude seed oils and oil fractions. J Agric Food Chem 51:6961–6969

Rao HSP, Reddy KS (1991) Isoflavones from *Flemingia vestita*. Fitoterapia 63:485

Rees SB, Harborne JB (1985) The role of sequiterpene lactones and phenolics in the chemical defence of the chicory plant. Phytochemistry 24:2225–2231. doi:0031-9422/85

Reuben DK, Aji SB, Andrew W, Abdulrahaman FI (2011) Preliminary phytochemical screening and in vitro anthelminticeEffects of aqueous extracts of *Salvadora persica* and *Terminalia avicennoides* against strongyline nematodes of small ruminants in Nigeria. J. Animal Vet Adv 10:437–442

Riou M, Guegnard F, Sizaret PY, Le Vern Y, Kerboeuf D (2010) Drug resistance is affected by colocalization of P-glycoproteins in raft-like structures unexpected in eggshells of the nematode *Haemonchus contortus*. Biochem Cell Biol 88:459–467. doi:10.1139/O09-1262760(95)00214-6

Rowan AD, Buttle DJ, Barrett AJ (1990) The cysteine proteinases of the pineapple plant. Biochem J 266:869–875

Roy B, Lalchhandama K, Dutta BK (2007) Anticestodal efficacy of *Accacia oxyphylla* on *Raillietina echinobothrida*: a light and electron microscopic studies. Pharmacologyonline 1:279–287

Sabah AA, Fletcher C, Webbe G, Doenhoff MJ (1986) *Schistosoma mansoni*: chemotherapy of infections of different ages. Exp Parasitol 61:294–303. doi:10.1016/0014-4894(86)90184-0

Sacchetti G, Maietti S, Muzzoli M, Scaglianti M, Manfredini S, Radice M, Bruni R (2005) Comparative evaluation of 11 essential oils of different origin as functional antioxidants, antiradicals and antimicrobials in food. Food Chem 91:621–632. doi:10.1016/j.foodchem.2004.06.031

Saeger B, Schmitt-Wrede HP, Dehnhardt M, Benten WPM, Krücken J, Harder A, von Samson-Himmelstjerna G, Wiegand H, Wunderlich F (2001) Latrophilin-like receptor from the parasitic nematode *Haemonchus contortus* as target for the anthelmintic depsipeptide PF1022A. FASEB J 15:1332–1334. doi:10.1096/fj.00-0664fje

Sahare KN, Anandhraman V, Meshram VG, Meshram SU, Reddy MVR, Tumane PM, Goswami K (2008) Anti-microfilarial activity of methanolic extract of *Vitex negundo* and *Aegle marmelos* and their phytochemical analysis. Indian J Exp Biol 46:128–131

Sakai C, Tomitsuka E, Esumi H, Harada S, Kita K (2012) Mitochondrial fumarate reductase as a target of chemotherapy: From parasites to cancer cells. Biochim Biophys Acta 1820:643–651. doi:10.1016/j.bbagen.2011.12.013

Salem ML (2005) Immunomodulatory and therapeutic properties of the *Nigella sativa* L. seed (review). Int Immunopharmacol 5:1749–1770. doi:10.1016/j.intimp.2005.06.008

Sánchez ME, Turina A del V, García DA, Nolan MV, Perillo MA (2004) Surface activity of thymol: implications for the eventual pharmacological activity. Colloids Surf B 34:77–86. doi:10.1016/j.colsurfb.2003.11.007

Sandoval-Castro CA, Torres-Acosta JFJ, Hoste H, Salem AZM, Chan-Pérez JI (2012) Using plant bioactive materials to control gastrointestinal tract helminths in livestock. Anim Feed Sci Tech 176:192–201. doi:org/10.1016/j.anifeedsci.2012.07.023

Sasaki T, Takagi M, Yaguchi T, Miyadoh S, Okada T, Koyama M (1992) A new anthelmintic cyclodepsipeptide, PF1022A. J Antibiot (Tokyo) 45:692–697

Satou T, Akao N, Matsuhashi R, Koike K, Fujita K, Nikaido T (2002a) Inhibitory effect of isoquinoline alkaloids on movement of second-stage larvae of *Toxocara canis*. Biol Pharm Bull 25:1651–1654

Satou T, Koga M, Matsuhashi R, Koike K, Tada I, Nikaido T (2002b) Assay of nematocidal activity of isoquinoline alkaloids using third-stage larvae of *Strongyloides ratti* and *S. venezuelensis*.Vet Parasitol 104:131–138. PII:S0304-4017(01)00619-7

Satou T, Horiuchi A, Akao N, Koike K, Fujita K, Nikaido T (2005) *Toxocara canis*: search for a potential drug amongst β-carboline alkaloids—in vitro and mouse studies. Exp Parasitol 110:134–139. doi:org/10.1016/j.exppara.2005.02.006

Satrija F, Nansen P, Bjorn H, Murtini S, He S (1994) Effect of papaya latex against *Ascaris suum* in naturally infected pigs. J Helminthol 68:343–346

Satrija F, Nansen P, Murtini S, He S (1995) Anthelmintic activity of papaya latex against patent *Heligmosomoides polygyrus* infections in mice. J Ethnopharmacol 48:161–164

Schweizer G, Braun U, Deplazes P, Torgerson PR (2005) Estimating the financial losses due to bovine fasciolosis in Switzerland. Vet Rec 157:188–193. doi:10.1136/vr.157.7.188

Scott JC, McManus DP (2000a) Molecular cloning and enzymatic expression of the 28-kDa glutathione S-transferase of *Schistosoma japonicum*: evidence for sequence variation but lack of consistent vaccine efficacy in the murine host. Parasitol Int 49:289–300. doi:S1383-5769

Scott JC, McManus DP (2000b) Molecular cloning and enzymatic expression of the 28-kDa glutathione S-transferase of *Schistosoma japonicum*: evidence for sequence variation but lack of consistent vaccine efficacy in the murine host. Parasitol Int 49: 289-300. doi:10.1016/S1383-5769(00)00058-1

Shakir L, Hussain M, Javeed A, Ashraf M, Riaz A (2011) Artemisinins and immune system. Eur J Pharmacol 668:6–14. doi:10.1016/j.ejphar.2011.06.044

Shalaby HA, Hatem AEl, Namaky AH, Kamel ROA (2009) In vitro effect of artemether and triclabendazole on adult *Fasciola gigantica*. Vet Parasitol 160:76–82. doi:10.1016/j.vetpar.2008.10.027

Silveira RX, Chagas ACS, Botura MB, Batatinha MJM, Katiki LM, Carvalho CO, Bevilaqua CML, Branco A, Machado EAA, Borges SL, Almeida MAO (2012) Action of sisal (*Agave sisalana*, Perrine) extract in the in vitro development of sheep and goat gastrointestinal nematodes. Exp Parasitol 131:162–168. doi:org/10.1016/j.exppara.2012.03.018

Singh TU, Kumar D, Tandan SK, Mishra SK (2009) Inhibitory effect of essential oils of *Allium sativum* and *Piper longum* on spontaneous muscular activity of liver fluke, *Fasciola gigantica*. Exp Parasitol 123:302–308. doi:10.1016/j.exppara.2009.08.002

Ghosh M, Sinha Babu, SP, Sukul NC, Mahato SB (1993) Antifilarial effect of two triterpenoid saponins isolated from *Acacia auriculiformis*. Indian J Exp Biol 31:604–606

Sinha Babu SP, Sarkar D, Ghosh NK, Saha A, Sukul NC, Bhattacharya S (1997) Enhancement of membrane damage by saponins isolated from *Acacia auriculiformis*. Jpn J Pharmacol 75:451–454

Sobhona P, Dangprasertc T, Chuanchaiyakuld S, Meepoola A, Khawsuka W, Wanichanona C, Viyanantb V, Upathamb ES (2000) *Fasciola gigantica*: ultrastructure of the adult tegument. Sci Asia 26:137–148

Socolsky C, Borkosky SA, Asakawa Y, Bardon A (2009) Molluscicidal phloroglucinols from the fern *Elaphoglossum piloselloides*. J Nat Prod 72:787–790. doi:10.1021/np800724hSSDI0378-874(95)01298-R

Soukhathammavong P, Odermatt P, Sayasone S, Vonghachack Y, Vounatsou P, Hatz CH, Akkhavong K, Keiser J (2011) Efficacy and safety of mefloquine, artesunate, mefloquine–artesunate, tribendimidine, and praziquantel in patients with *Opisthorchis viverrini*: a randomised, exploratory, open-label, phase 2 trial. Lancet Infect Dis 11:110–118. doi:10.1016/S1473-3099(10)70250-4

Spicher M, Roethlisberger C, Lany C, Stadelmann B, Keiser J, Ortega-Mora LM, Gottstein B, Hemphill A (2008) In Vitro and in vivo treatments of *Echinococcus* protoscoleces and metacestodes with artemisinin and artemisinin derivatives. Antimicrob Agent Chemoth 2:3447–3450. doi:10.1128/AAC.00553-08

Squires JM, Ferreira JFS, Lindsay DS, Zajac AM (2011) Effects of artemisinin and *Artemisia* extracts on *Haemonchus contortus* in gerbils (*Meriones unguiculatus*). Vet Parasitol 175:103–108. doi:10.1016/j.vetpar.2010.09.011

Stepek G, Behnke JM, Buttle DJ, Ducel IR (2004) Natural plant cysteine proteinases as anthelmintics? Trends Parasitol 20:322–327. doi:10.1016/j.pt.2004.05.003

Stepek G, Lowe AE, Buttle DJ, Duce IR, Behnke JM (2006) In vitro and in vivo anthelmintic efficacy of plant cysteine proteinases against the rodent gastrointestinal nematode, *Trichuris muris*. Parasitology 132:681–689. doi:10.1017/S003118200500973X

Stepek G, Lowe AE, Buttle DJ, Duce IR, Behnke JM (2007a) Anthelmintic action of plant cysteine proteinases against the rodent stomach nematode, *Protospirura muricola*, in vitro and in vivo. Parasitology 134:103–112. doi:10.1017/S0031182006001302

Stepek G, Lowe AE, Buttle DJ, Duce IR, Behnke JM (2007b) In vitro anthelmintic effects of cysteine proteinases from plants against intestinal helminths of rodents. J Helminthol 81:353–360. doi:10.1017/S0022149X0786408X

Stepek G, Lowe AE, Buttle DJ, Duce IR, Behnke JM (2007c) The anthelmintic efficacy of plant-derived cysteine proteinases against the rodent gastrointestinal nematode, *Heligmosomoides polygyrus*, in vivo. Parasitology 134:1409–1419. doi:10.1017/S0031182007002867

Tadros MM, Ghaly NS, Moharib MN (2008) Molluscicidal and schistosomicidal activities of a steroidal saponin containing fraction from *Dracaena fragrans* (L.). J Egypt Soc Parasitol 38:585–598

Takahashi Y, Matsumoto A, Seino A, Ueno J, Iwai Y, Ōmura S (2002) *Streptomyces avermectinius* sp. nov., an avermectin-producing strain. Int J Syst Evol Microbiol 52:2163–2168. doi:10.1099/ijs.0.02237-0

Takano D, Nagamitsu T, Ui H, Shiomi K, Yamaguchi Y, Masuma R, Kuwajima I, Oʼmura S (2001) Absolute configuration of nafuredin, a new specific NADH-fumarate reductase inhibitor. Tetrahedron Lett 42:3017–3020. doi:PII:S0040-4039(01)355-0

Tandon V, Pal P, Roy B, Rao HSP, Reddy KS (1997) In vitro anthelmintic activity of root tuber extract of *Flemingia vestita*, an indigenous plant in India. Parasitol Res 83:492–498

Tandon V, Das B, Saha N (2003) Anthelmintic efficacy of *Flemingia vestita* (Fabaceae): effect of genistein on glycogen metabolism in the cestode, *Raillietina echinobothrida*. Parasite Int 52:179–183. doi:org/10.1016/S1383-5769(03)00006-0

Tangpu VT, Yadav AK (2006) Anticestodal property of *Strobilanthes discolor*: an experimental study in *Hymenolepis diminuta*—rat model. J Ethnopharmacol 105:459–463. doi:10.1016/j.jep.2005.11.015

Tangpu VT, Temjenmongla K, Yadav AK (2004) Anticestodal activity of *Trifolium repens* extracts. Pharmaceut Biol 42:656–658. doi:10.1080/13880200490902617

Tansatit T, Sahaphong S, Riengrojpitak S, Viyanant V, Sobhon P (2012) *Fasciola gigantica*: the in vitro effects of artesunate as compared to triclabendazole on the 3-weeks-old juvenile. Exp Parasitol 131:8–19. doi:org/10.1016/j.exppara.2012.02.018

Tariq KA, Chishti MZ, Ahmad F, Shawl AS (2009) Anthelmintic activity of extracts of *Artemisia absinthium* against ovine nematodes. Vet Parasitol 160:83–88. doi:10.1016/j.vetpar.2008. 10.084

Thomas TRA, Kavlelar DP, LokaBharathi PA (2010) Marine drugs from sponge-microbe association—a review. Mar Drugs 8:1417–1468. doi:10.3390/md8041417

Turina AV, Nolan MV, Zygadlo JA, Perillo (2006) Natural terpens: self-assebly and membrane partitioning. Biophys Chem 122:101–113. doi:10.1016/j.bpc.2006.02.007

Türkdoğan MK, Ağaoğlu Z, Yener Z, Sekeroğlu R, Akkan HA, Avci ME (2001) The role of antioxidant vitamins (C and E), selenium and *Nigella sativa* in the prevention of liver fibrosis and cirrhosis in rabbits: new hopes. Dtsch Tierarztl Wochenschr 108:71–73

Tzamaloukas O, Athanasiadou S, Kyriazakis I, Jackson F, Coop RL (2005) The consequences of short-term grazing of bioactive forages on established adult and incoming larva populations of *Teladorsagia circumcincta* in lambs. Int J Parasitol 35:329–335. doi:10.1016/ j.ijpara.2004.11.013

Tzamaloukas O, Athanasiadou S, Kyriazakis I, Huntley JF (2006) The effect of chicory (*Cichorium intybus*) and sulla (*Hedysarum coronarium*) on larval development and mucosal cell responses of growing lambs challenged with *Teladorsagia circumcincta*. Parasitology 132:419–426. doi:10.1017/S0031182005009194

Utzinger J, Xiao SH, Goran EKN, Bergquist R, Tanner M (2001) The potential of artemether for the control of schistosomiasis. Int J Parasitol 31:1549–1562. doi:10.1016/S0020-7519(00)00297-1

Utzinger J, Xiao SH, Tanner M, Keiser J (2007) Artemisinins for schistosomiasisand beyond. Curr Opin Investig Drugs 8:105–116

Utzinger J, Raso G, Brooker S, DeSavigny D, Tanner M, Ørnbjerg N, Singer BH, Goran EKN (2009) Schistosomiasis and neglected tropical diseases: towards integrated and sustainable control and a word of caution. Parasitology 136:1859–1874. doi:10.1017/ S0031182009991600

Várady M, Čorba J, Letková V, Kováč G (2009) Comparison of two versions of larval development test to detect anthelmintic resistance in *Haemonchus contortus*. Vet Parasilot 160:267–271. doi:10.1016/j.vetpar.2008.11.010

Varinska L, Mirossay L, Mojzisova G, Mojzis J (2010) Antiangogenic effect of selected phytochemicals. Pharmazie 65:57–63. doi:10.1691/ph.2010.9667

Verdrengh M, Collins LV, Bergin P, Tarkowski A (2004) Phytoestrogen genistein as an anti-staphylococcal agent. Microbes Infect 6:86–92. doi:10.1016/j.micinf.2003.10.005

von Son-de Fernex E, Alonso-Díaz MA, Valles-de la Mora B, Capetillo-Leal CM (2012) In vitro anthelmintic activity of five tropical legumes on the exsheathment and motility of *Haemonchus contortus* infective larvae. Exp Parasitol 131:413–418. doi:org/10.1016/ j.exppara.2012.05.010

Vuong D, Capon RJ, Lacey E, Gill JH, Heiland K, Friedel T (2001) Onnamide F: a new nematocide from a southern Australian marine sponge, *Trachycladus laevispirulifer*. J Nat Prod 64:640–642. doi:10.1021/np000474b

Waghorn GC, McNabb WC (2003) Consequences of plant phenolic compounds for productivity and health of ruminants. Proc Nutr Soc 62:383–392. doi:org/10.1079/PNS2003245

Waller PJ (2006) From discovery to development: current industry perspectives for the development of novel methods of helminth control in livestock. Vet Parasitol 139:1–14. doi:10.1016/j.vetpar.2006.02.036

Waterman PG (1999) The tannins - an overview. In Brooker JD (ed) Tannins in livestock and human nutrition. Proceedings of international workshop, Adelaide, Australia, Australian Centre for International Agricultural Research, pp 10–13. doi: 10.1016/j.pt.2006.04.004

Watson M (2009) Praziquantel. Review. J Exotic Pet Med 18:229–231. doi:10.1053/
    j.jepm.2009.06.005
Watts KR, Tenney K, Crews P (2010) The structural diversity and promise of antiparasitic marine
    invertebrate-derived small molecules. Curr Opinion Biotechnol 21:808–818. doi:10.1016/
    j.copbio.2010.09.015
Weissenberg M (2001) Isolation of solasodine and other steroidal alkaloids and sapogenins by
    direct hydrolysis-extraction of *Solanum* plants or glycosides therefrom. Phytochemistry
    58:501–508. doi:10.1016/S0031-9422(01)00185-6
Wu LJ, Li SW, Xuan YX, Xu PS, Liu ZD, Hu LS, Zhou SY, Qiu YX, Liu YM (1995) Field
    application of artesunate in prophylaxis of schistosomiasis: an observation of 346 cases. Chin
    J Schisto Control 7:323–327 (in Chinese)
Xiao SH (2005a) Development of antischistosomal drugs in China, with particular consideration
    to praziquantel and the artemisinins. Acta Trop 96:153–167. doi:10.1016/j.actatropica.
    2005.07.010
Xiao SH (2005b) Study on prevention and cure of artemether against schistosomiasis. Chin J
    Schisto Control 17:310–320 (in Chinese)
Xiao SH, Catto BA (1989) In vitro and in vivo studies of the effect of artemether on *Schistosoma
    mansoni*. Antimicrob Agent Chemother 33:1557–1562. doi:10.101128/AAC.33.9.1557
Xiao SH, Yue WJ, Yang YQ, You JQ (1987) Susceptibility of *Schistosoma japonicum* to different
    developmental stages to praziquantel. Chin Med J 100:759–768
Xiao SH, You JQ, Yang YQ, Wang CZ (1995) Experimental studies on early treatment of
    schistosomal infection with artemether. Southeast Asian J Trop Med Public Health 26:306–
    318
Xiao SH, Hotez PJ, Tanner M (2000a) Artemether, an effective new agent for chemoprophylaxis
    against schistosomiasis in China: its in vivo effect on the biochemical metabolism of the
    Asian schistosome. Southeast Asian J Trop Med Public Health 31:724–732
Xiao SH, Chollet J, Weiss NA, Bergquist RN, Tanner M (2000b) Preventive effect of artemether
    in experimental animals infected with *Schistosoma mansoni*. Parasitol Int 49:19–24
Xiao SH, Utzinger J, Chollet J, Endriss Y, N'Goran EK, Tanner M (2000f) Effect of artemether
    against *Schistosoma haematobium* in experimentally infected hamsters. Int J Parasitol
    30:1001–1006. doi:10.1016/S0020-7519(00)00091-6
Xiao SH, Ji-Qing Y, Hui-Fang G, Jin-Yan M, Pei-Ying J, Chollet J, Tanner M, Utzinger (2002)
    Schistosoma japonicum: effect of artemether on glutathione S-transferase and superoxide
    dismutase. Exp Parasitol 102:38–45. doi:10.1016/S0014-4894(02)00145-5
Xiao SH, Xue J, Tanner M, Zhang Yong-Nian, Keiser J, Utzinger J, Qiang H-Q (2008)
    Artemether, artesunate, praziquantel and tribendimidine administered singly at different
    dosages against *Clonorchis sinensis*: a comparative in vivo study. Acta Tropica 106: 54–59.
    doi:10.1016/j.actatropica.2008.01.003
Xiao SH, Keiser J, Xue J, Tanner M, Morson G, Utzinger J (2009) Effect of single-dose oral
    artemether and tribendimidine on the tegument of adult *Clonorchis sinensis* in rats. Parasitol
    Res 104:533–541. doi:10.1007/s00436-008-1227-6
Xiao SH, Keiser J, Chen MG, Tanner M, Utzinger J. (2010) Research and development of
    antischistosomal drugs in the People's Republic of China: a 60-year review. In: Zhou XN,
    Bergquist R, Olveda R et al (eds) Adv Parasitol 73:231–295. doi:10.1016/S0065/S0065-
    308X(10)73009-8
Yadav AK, Tangpu V (2008) Anticestodal activity of *Adhatoda vasica* extracts against
    Hymenolepis diminuta infection in rats. J Ethnopharmacol 119:322–324. doi:10.1016/
    j.jep.2008.07.012
Yadav AK, Tangpu V (2009) Therapeutic efficacy of *Zanthoxylum rhetsa* DC extract against
    experimental *Hymenolepis diminuta* (Cestoda) infections in rats. J Parasitol Dis 33:42–47.
    doi:10.1007/s12639-009-0007-2
Yadav AK, Tangpu V (2012) Anthelmintic activity of ripe fruit extract of *Solanum myriacanthum
    Dunal* (Solanaceae) against experimentally induced *Hymenolepis diminuta* (Cestoda)
    infections in rats. Parasitol Res 110:1047–1053. doi:10.1007/s00436-011-2596-9

Yang YQ, Xiao SH, Tanner M, Utzinger J, Chollet J, Wu JD, Guo J (2001) Histopathological changes in juvenile *Schistosoma haematobium* harboured in hamsters treated with artemether. Acta Trop 79:135–141. doi:10.1016/S0001-706X(01)00069-9

Zang X, Maizels RM (2001) Serine proteinase inhibitors from nematodes and the arms race between host and pathogen. Trends Biochem Sci 26:191–197. doi:PII:S0968-0004(00)01761-8

Zibaei M, Sarlak A, Delfa B, Ezatpour B, Azargoon A (2012) Scolicidal effects of *Olea europaea* and *Satureja khuzestanica* extracts on protoscolices of hydatid cysts. Korean J Parasitol 50:53–56. doi:org/10.3347/kjp.2012.50.1.53

Ziegler J, Facchini PJ (2008) Alkaloid biosynthesis: metabolism and trafficking. Ann Rev Plant Biol 59:735–769. doi:10.1146/annurev.arplant.59.032607.092730

Chapter 3
Natural Compounds Exerting Anthelmintic and/or Host-Protecting Effects During Parasitic Infections

Abstract Helminth parasites are able to regulate the host's defense mechanisms in order to prevent their expulsion or killing and these are characterized by chronic immunosuppression and reduced pathology. Hence, any combined therapy would take advantage of the synergistic action of a drug and a natural compound, which exert immunomodulatory as well as antiparasitic activities. To date, only a few natural compounds with defined molecular structure and well-described biological and medicinal activities have been investigated as having also an anthelmintic effect. In this chapter, after a brief introduction to key features of immunosuppression and host pathology, anthelmintic and immunomodulatory activities of artemisinins, genistein, curcumin, and tannins are described with some insight into selective toxicity to pathogens and cancer cells, but very low toxicity to the normal cells in the hosts. The host-protecting effect of a natural compound with antioxidant and antifibrotic activities in reducing the host pathology, thus, contributing to the elevated drug efficacy, is also highlighted in this chapter. The focus is on phenolic compounds paeoniflorin and silymarin, so far examined in flatworm infections, and the description of several molecular mechanisms underlying the above beneficial effects is provided. Regarding immunomodulatory activity of all phenolic compounds reviewed, the direction to selected subsets of immune cells toward the immunological balance depends on the type of disease.

Keywords Helminths · Natural phenolic compounds · Immunosuppression · Pathology · Anthelmintic activity · Artemisinins · Genistein · Curcumin · Tannins · Paeoniflorin · Silymarin

G. Hrckova and S. Velebny, *Pharmacological Potential of Selected Natural Compounds in the Control of Parasitic Diseases*, SpringerBriefs in Pharmaceutical Science & Drug Development, DOI: 10.1007/978-3-7091-1325-7_3, © The Author(s) 2013

3.1 Host Pathology and Immunosuppression During Helminth Infections

Parasite infection and its corresponding host immune response are a result of long-term coevolution. It is a disadvantage for the parasite to kill its host; rather it deceives the host into developing an ineffective immune response. The parasites that have survived during evolution are well adapted to their host and show marked host specificity. Helminths are masterful immunoregulators of vertebrate host defense systems, which they can evade in order to complete the life cycle or persist in the host for prolonged periods; therefore, infections generally have a chronic character. The immune response against helminths includes both innate and adaptive components where innate immune cells are essential for initiation of defense mechanism against the metazoan. The adaptive components instruct and amplify the innate effector-cell response primarily through the secretion of cytokines (Anthony et al. 2007). The following brief description of the main immunological and pathological characteristics of helminth infections was aimed to provide some background to issue how selected natural compounds could be beneficial for the host by enhancing immune responses and attenuating the pathology.

Characteristic features of helminth infections are hypergammaglobulinemia, with the IgE class of antibodies being typical and IgG as the most abundant class. A typical characteristic trait of helminth infections is dominance of Th2-type immune response characterized by Th2 related cytokines that include IL-4, IL-5, IL-13, and IL-10. These cytokines induce B-lymphocytes to switch to IgE antibody production, promote alternative macrophage activation, eosinophil maturation and recruitment, and stimulation of immunoregulatory cell populations, such as regulatory T cells (Hewitson et al. 2009). Immunosuppression of effector functions is a well-described phenomenon during chronic infections. The initial exposure to allergen or parasite antigen leads to activation of proinflammatory Th1 type which is quickly switched to T helper 2 (Th2) cells. They orchestrate the immune response during the chronic stage of diseases through secretion of the above-mentioned cytokines. Additionally, the trafficking of eosinophils into inflammatory sites involves the chemokine interactions (e.g., eotaxin), lipids (e.g., LTB4), and adhesion molecules (e.g., P-selectin) (Rothenberg and Hogan 2006).

The regulatory network associated with chronic helminth infections persisting in the host tissues, is believed to prevent strong immune responses against parasitic worms, allowing the long-term survival and restricting pathology in the host. The broad topic dealing with mechanisms and consequences of immunomodulation in various helminth infections has been reviewed by several authors (for example: (Maizels and Yazdanbakhsh 2003; van Riet et al. 2007; Gottstein and Hemphill 2008). It was documented in a numerous studies and reviews that immunomodulatory effects are associated with soluble mediators, which ligate, degrade, or otherwise interact with host immune cells and molecules (for example see reviews: Lightowlers and Rickard 1988; Freitas and Pearce 2010). With the development of helminth genomics, systematic proteomic analyses of many major helminth excretory–secretory products revealed that a common set of proteins secreted by helminths includes proteases, protease inhibitors, venom allergen homologs, glycolytic enzymes, and lectins (Hewitson et al. 2009). Such an immunomodulation by tissue helminths is hypothesized to be beneficial to both the mammalian hosts and the parasite, as it could protect helminths from being killed, and at the same time protect the host from excessive proinflammatory responses (Th1 type) that may lead to organ damage. Pathological consequences involve oxidative damage to the normal host cells and stimulation of a cascade of profibrotic factors which lead to tissue fibrogenesis and formation of fibrous granulomas or capsules around the parasites. Extensive studies of several health threatening diseases in humans (for example: schistosomiasis, echinococcosis) affecting liver and other parenchymal organs have revealed that granuloma formation is attributable to CD4+Th2-driven response and delayed type of hypersensitivity (for example: Burke et al. 2009; Dixon 1997). Encapsulation of parasites in granulomas and fibrous capsules is a host-protecting response and such a barrier should prevent migration of parasites or limit diffusion of antigens secreted by eggs or larvae to surrounding tissue. On the other hand, fibrosis and granulomas decrease the bioavailability of anthelmintic drugs for entrapped parasites/eggs. Among other consequences of chronic infections are the presence of circulating antigens, persistent antigenic stimulation, and the formation of immune complexes. The formation of immune complexes and their localization in the tissues or blood may give rise to many pathological effects, for example kidney problems or induction of allergy-type symptoms associated with eosinophilia, which is a hallmark of allergic and parasite diseases.

Although all helminth infections are characterized by their ability to induce Th2-cell responses and antibody-mediated cytotoxic activity of immune cells, immunity during gastrointestinal infections differs in some specific features from immune responses induced by the tissue-dwelling developmental stages of helminths. Generally, Th2-cell responses elicited by gastrointestinal infections result in eosinophilia, goblet, and mucosal mast-cell hyperplasia, and the production of noncomplement fixing antibodies; however, immune response differs considerably between different helminths. The review by Gause et al. (2003) specifically focussed on the factors that regulate the immune responses and the immunopathology in certain, well-studied intestinal nematode parasites in monogastric animal models. Host–parasite interactions via Th2 type responses in the gastrointestinal tract of the host are directed to prevent expulsion of adult worms, which release a high number of eggs into the feces of the host.

Thus, Th2-type response exemplifies processes which lead to the worm expulsion and on the other hand it restricts pathological inflammation, via downregulation of the acute stage Th1 type of immune response (Anthony et al. 2007). Immunosuppression is advantageous for parasites and reduces the ability of host to eliminate well-adapted parasites. Therefore, natural compounds which would be able to stimulate suppressed effect or immune mechanisms, as well as, to act synergistically with anthelmintic drugs, can be proposed as an effective alternative therapy. In respect of the above-mentioned pathological consequences during chronic infections, coadministration of a natural compound which would act as an antifibrotic, antioxidant, and/or immunostimulating agents, seems to be another promising treatment alternative.

3.2 Compounds Affecting Parasites and Mammalian Hosts: Anthelmintic Action Versus Benefit for Hosts

The majority of plants with a record in ethnomedicine as being effective remedies against parasitic infections in humans and animals, are believed to heal also many other nonparasitic diseases. Plants as well as marine organisms are known to produce a large variety of small molecules of different chemical natures. An interesting observation was that most of these small molecules have weak antibiotic activity—several orders of magnitudes less than that of common antibiotics produced by bacteria and fungi. In spite of the fact that plant-derived antibacterials are less potent, plants fight infections successfully. Hence, it becomes apparent that plants adopt a different paradigm—"synergy"—to combat infections (Hemaiswarya et al. 2008). This lesson given by plant compounds could be adopted as the strategy for a more effective antiparasitic therapy. The combined therapy would take an advantage of synergistic action of a drug and a natural compound which exert antiparasitic as well as immunomodulatory activities. It has been demonstrated by numerous studies reviewed in this book that many isolated plant secondary metabolites possess direct anthelmintic activity and there is a large body of evidence that most metabolites are capable of directly affecting inflammatory mediators, as well as the production/activity of second messengers, transcriptions factors, and key proinflammatory molecules (see for review: Calixto et al. 2003, 2004). Some compounds, which showed direct antiparasitic effects in vitro and in animal models, have been widely examined for their broad therapeutical potential in non-parasitic diseases. The list of such natural compounds should include *artemisinin and its derivates, curcumin, genistein, and tannins* (Fig. 3.1). The use of multiple bioassays in pharmacological testing is important as it gives a clearer indication of the effect of the extracts in relation to the disease state. In addition, it reduces the possibility of losing other potentially useful bioactive compounds present in the investigated plant extracts (Houghton et al. 2007). The evaluation of all the investigational new drugs for their potential immunomodulatory effects such as immunosuppression, immunomodulation, autoimmunity, hypersensitivity, etc., is an obligate recommendation for Investigational New Drug (IND) procedures of the FDA. This obligatory exercise also focuses on clinically available classes, being extremely helpful in assessing the extent of a drug's overall therapeutic effect and toxicity (Shakir et al. 2011).

Artemisinin is a unique secondary plant metabolite in terms of its chemical structure as it contains an endoperoxide bridge in the molecule. This sesquiterpene lactone is present in a high amount in *Artemisia annua* and indication of strong

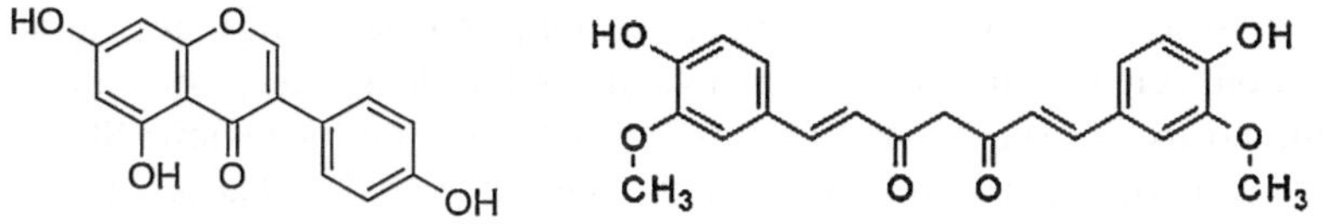

Genistein   Curcumin (Keto form)

Fig. 3.1 Chemical structure of curcumin and genistein

antiprotozoal activity came from Chinese herbal medicine. The mechanisms of action of artemisinins on parasitic protozoa, pharmacological properties, toxicity, and clinical applications have been subjected to numerous studies and reviewed by several authors, for example: Balint (2001), Meschnik (2002), Golenser et al. (2006), Olliaro et al. (2001), and the most relevant information about this issue is summarized in Chap. 1.

The maintenance of calcium ion homeostasis is vital for all organisms. While low intracellular concentrations of the ion are required for signal transduction and for the active sites of some enzymes, higher concentrations can cause cell death (Duchen 2000). Therefore, organisms must precisely regulate the concentration of calcium ions in their cells. Eukaryotes achieve this partly through the active transport of calcium ions across cellular membranes. One group of proteins responsible for this task is calcium ATPases. This family of transmembrane, ATP-dependent calcium ion pumps, can be subdivided into three main classes, the sarco(endo)plasmic reticulum Ca^{2+}-ATPases (SERCA), the secretory pathway Ca^{2+}-ATPases (SPCA), and the plasma membrane Ca^{2+}-ATPases (PMCA) (Vanoevelen et al. 2005). The calcium ATPases are targets for a number of drugs and artemisinins are believed to target a specific SERCA isoform in protozoan parasites (Eckstein-Ludwig et al. 2003). It is not known whether Ca^{2+}-ATPases in metazoan parasites are affected by artemisinins.

Here, we wish to present artemisinins not only as a potent drug for the therapy of trematode infections (see Chap. 2) in humans but also as a promising active principle against cancer cells and potent immunomodulators. Metabolism of drugs, although often needed for their activation, sometimes leads to their inactivation. While artemisinin is converted primarily into inactive metabolites, its derivatives artesunate, artemether, and arteether are metabolized to the clinically more effective dihydroartemisinin (DHA), which has about six time longer life in the circulation than parent compounds (Balint 2001). Combination therapies using artemisinin derivates and antiprotozoal drugs, which have another mechanism of action and longer elimination half-life, are currently recommended by the World Health Organization (WHO) as the first-line antimalarial treatment for *Plasmodium falciparum* (World Health Organization 2005). A combined therapy with artemether/artesunate and praziquantel was recommended also for schistosomiasis and trematodiasis as a significantly more effective treatment strategy (Keiser and Utzinger 2007; Liu et al. 2011; Keiser et al. 2010).

Since being recognized as effective antiprotozoal drugs for large-scale application, artemisinins have been evaluated for their effects on the immune system, mostly in experimental protozoal infections (see for review: Shakir et al. 2011). Numerous studies reviewed by the authors revealed that artemisinins are armed with a versatile immunomodulatory impact on various elements of the immune system. For instance, they demonstrated predominant immunosuppressive traits toward different immune components by particularly regulating the cellular proliferation and cytokine release, which indicates that they possess some additional mechanisms and features. Summarized is also information about immuno-modulatory effects of artemisinins on different immune cells including neutrophils, macrophages, splenocytes, and T and B cells in conjunction with their therapeutic prospective with regard to inflammation, autoimmunity, and delayed-type hyper-sensitivity. Induction of cell apoptosis by artemisinin was also demonstrated (Zhao et al. 2007).

Many conclusions regarding immunosuppressive effects are controversial and discrepancies seem to be related to a disease-specific immunity. Macrophages have a double role in cell-mediated (act as antigen presenting and effector cells) as well as in humoral immunity. They secrete variety of pro-inflammatory cytokines by NF-kB-mediated pathway during acute and chronic inflammation (Fujiwara and Kobayashi 2005). Artemisinins can diminish the secretion of macrophage-derived proinflammatory cytokines, particularly of TNF and IL-1β via inhibition of NF-kB in human adherent monocytes (Wang et al. 2006). In addition, artemisinin (50 mg/kg, intraperitoneal) has depicted strong anti-inflammatory activity in rats suffering from acute pancreatitis.

A notable discrepancy, however, has been observed recently related to arte-misinin regarding iNOS mRNA expression and NO production. On the one hand, artemisinin has been reported to suppress iNOS mRNA expression and NO production in LPS-activated macrophages, which was observed to be due to inhibition of IFN-β production (Kang et al. 2010). On the other hand, another study has ascertained that artemisinin can induce a host protective response by renormalization of attenuated nitrite production as well as iNOS mRNA expres-sion in *Leishmania donovani*-infected macrophages (Sen et al. 2010). Artemisinins may also interfere with functional competencies of neutrophils. In view of the above, it is highly possible that immunomodulatory effect of this compound might have a significant impact on the outcome of therapy of human metazoan infection, namely schistosomiasis and fascioliasis.

To show the full potential of artemisinins, their potent and broad anticancer properties in cell lines and animal models should be mentioned and are described in the review of Krishna et al. (2008). Artemisinins such as artesunate were found to be active against a variety of unrelated tumor cells lines, from the most common types such as colon, breast, and lung cancers to leukemias and pancreatic cancer (Efferth et al. 2001). One of the best described pathways for anticancer activity is inhibition of enhanced angiogenesis associated with tumors. Artesunate is a cheap, safe, easily administered, and orally bioavailable compound that acts on targets different from those of many current cancer chemotherapeutic agents and is unlikely to interact adversely with existing anticancer interventions.

Curcumin, a polyphenolic organic molecule which is the major constituent of *Curcuma longa*, is an interesting molecule and is an example of compound acting on several biological systems (Araujo and Leon 2001). Antiparasitic effects of curcuma have been reviewed recently (Haddad et al. 2011) and were demonstrated also against *Plasmodium* (Mishra et al. 2008) and *L. donovani* (Das et al. 2008). In Sect. 2.3 we reviewed its direct toxic effect on *Schistosoma mansoni* adults, which was manifested as the disruption of worm tegument at low micromolar concentrations (Magalhães et al. 2009). However, its molecular target in metazoan parasites has not been revealed to date, but some attempts have already been made. In a novel high-throughput fluorometric screening assay curcumin did not inhibit SmNACE up to 100 μM (Kuhn et al. 2010). In this context, the newly characterized *Schistosoma mansoni* NAD$^+$ catabolizing enzyme (SmNACE) represents a potentially attractive drug target. This potent NAD$^+$glycohydrolase, which is localized to the outer surface (tegument) of the adult parasite, is presumably involved in the parasite survival by manipulating the host's immune regulatory pathways.

Curcumin also has many therapeutic applications reviewed, for example, by Aggarwal and Harikumar (2009). They include its much investigated anticancer activity (Agrawal and Mishra 2010) as well as antiinflammatory, antioxidant, and antiviral activities (Maheshwari et al. 2006). Regarding its cytotoxicity, in vivo studies with curcumin showed lack of significant toxicity (Perkins et al. 2002). Modulation of immune responses in the course of murine schistosomiasis following administration of curcumin has been demonstrated in the study of Allam (2009). Apart from a significant reduction of worm burden and tissue-egg burdens, curcumin treatment reduces granuloma volume by 79 %, restored hepatic enzyme activities to the normal levels and enhanced catalase activity in the liver tissue of infected mice. Moreover, hepato-splenomegaly and eosinophilia induced by *S. mansoni* infection were largely improved with curcumin treatment. Mice treated with curcumin showed low serum levels of both interleukin (IL)-12 and tumor necrosis factor alpha (TNF-α), but IL-10 level was not significantly altered. Specific IgG and IgG1 responses against both soluble worm antigen and soluble egg antigen (SEA) were augmented with curcumin treatment, but IgM and IgG2a responses were not significantly changed. In view of the reported in vitro antiparasitic activity and beneficial immunomodulatory and antifibrotic effects, curcumin could be suggested as the suitable adjunct to therapy of schistosomiasis, and also other chronic infections with similar immunopathology.

The isoflavone *genistein* has been shown to exert beneficial effects on many disorders, including cancer and cardiovascular diseases and has been known for a long time as a phytoestrogen (for example: Polkowski and Mazurek 2000; Taylor et al. 2009). Antibacterial activity was also confirmed, for example, exposure to genistein exhibited an inhibitory effect on all staphylococcal strains tested (Verdrengh et al. 2004). The great interest that has focused on genistein led to the identification of numerous intracellular targets in mammalian cells. At the molecular level, it inhibits the activity of ATP utilizing enzymes such topoisomerase II, tyrosine-specific protein kinase, and enzymes involved in phosphatidylinositol turnover. It was suggested to beneficially affect lipid metabolism in humans and thereby it can contribute to healthy aging. At the cellular level, genistein induces apoptosis and inhibits differentiation in cancer cells as well as their proliferation. It can act as antioxidant agent and was used as immunosuppressive agent both in vitro and in vivo. The role of nitric oxide (NO) production by activated macrophages in inflammatory lesions surrounding parasites was recognized to contribute to more severe pathology. In this respect, genistein and other isoflavones found in plant extracts and exerting anthelmintic activity (for example: kaempherol, quercetin), inhibited activation of nuclear factor κB (NF- κB), which is a significant transcription factor for inducible nitric oxide synthase (iNOS) (Hämäläinen et al. 2007). The direction of genistein activity to selected subsets of immune cells depends on the type of disease and it tends to alter Th1/Th2 balance. In line with this hypothesis, it significantly suppressed the secretion of IFN-γ and augmented the IL-4 production in a rheumatoid arthritis model (Wang et al. 2008), but it decreased Th2-type cytokines, including IL-4, in murine asthma model (Gao et al. 2012)

In vitro activity of genistein isolated from root-tuber extract of the plant *Flemingia vestita* has been evaluated on several helminths and data have been summarized in Chap. 2. Information obtained from studies on a trematode model (*Fasciolipsis bruski* –adult) and cestode models (*Raillientina echinobothrida*-adult, *Echinococcus multilocularis* and *E. granulosus*–larval stage) demonstrated that genistein is an effective trematocidal and cestocidal molecule as it was able to inhibit several essential physiological mechanisms in these flatworms. However, its therapeutical potential in controlled in vivo experiments has to be examined. In contrast with susceptibility of flatworms, genistein was not an effective nematocidal compound in experimental studies on *Ascaris suum* from pigs, *A. lumbricoides* from humans, *Ascaridia galli* and *Heterakis gallinarum* from domestic fowl (Tandon et al. 1997). In this in vitro study, nematodes did not show any change in physical activity and remained viable even after a long period of exposure to the extract. The free–living nematode *Caenorhabditis elegans* was employed in a few studies designed to explain the ineffectiveness of genistein toward nematodes. Fischer et al. (2012) showed that this isoflavone reduced resistance to nematode pathogen *Photorhabdus luminescens* via reduction of vitellogenin-expression, which are invertebrate estrogen-responsive genes. Previous studies on flatworms demonstrated that a putative molecular target for genistein is localized in the nervous system and is involved in coordination of worm motility. The function of genistein as a neuronal mediator was evaluated on the human α7 nicotinic acetylcholine receptor (nAChR) subunit and its *C. elegans* homologue, ACR-16. It is well-known that acetylcholine receptors and acetylcholine-regulated processes are vital for humans and helminths, and in nematodes they were shown to be associated with resistance to benzimidazole carbamate anthelmintics (Sutherland and Lee 1993). Functional recombinant homomeric receptors were generated and expressed in *Xenopus laevis* oocytes but genistein did not show any positive modulation of ACR-16 receptor subunit in contrast with marked effect on human receptor (Sattelle et al. 2009). To our knowledge, an immunomodulatory effect of genistein during experimental in vivo studies in hosts with flatworm infections was not confirmed. Due to its broad biological activities in humans, genistein interactions with the immune cells were subjected to an immense research. In allergic types of disease, such murine asthma models induced by sensitization with ovalbumin, in vivo administration of genistein was examined. This flavonoid attenuated inflammation, decreased Th2-type cytokines production by CD4+lymphocytes in the favor of Th1 type of cytokines (IFN-γ) (Gao et al. 2012). Immunity during helminth infections is also characterized by Th2-dominant cell and cytokine responses; therefore, therapy with well-tailored doses of genistein would benefit from its dual effect, vermicidal against flatworm infections as well as immunomodulatory leading to the activation of an antiparasitic Th1 type of response.

Tannins are polyphenolic substances often discussed as effective compounds in natural control of gastrointestinal nematode infections in livestock, mainly sheep and goats. Several isolated tannins have been shown to have antibacterial and antiprotozoal activities and/or to stimulate non-specific immune systems. Hydrolyzable tannins formed of flavan-3-ols with galloyl groups showed immunomodulatory effects on murine macrophage cells lines, markedly stimulated release of TNF but release of NO was only moderately changed (Kolodziej et al. 2001). Plants of genus *Anacardium* are rich in secondary metabolites with a high portion of tannins. The anti-nematode activity against GIN of sheep in vitro and/or in vivo was found in extracts of *Anacardium humile* containing mainly tannins, flavonoids, and alkaloids with $LD_{50} = 10.14$ mg/ml (Nery et al. 2010). Hydrolysable and condensed tannins obtained from the bark of *A. occidentale* L. demonstrated apparent anti-inflammatory activity on non-parasitic disease model in vivo (Mota et al. 1985). In this study tannins administered to rats intraperitoneally were found to inhibit some inflammatory mediators responsible for elevation of permeability of blood vessels which was demonstrated by the strong inhibition of oedema formation in rats. Moreover, tannins administered daily for 7 days in doses of 6.25–50 mg/kg produced a dose-dependent reduction of the granuloma tissue formation induced chemically as well as the inhibition of migration of leucocytes to these sites. It seems that tannins may produce effects in a non-specific manner by their astringent properties on cell membranes, thus affecting cell functions. Condensed tannins (proanthocyanidins) are present in higher concentrations in plants used as forage for livestock and were shown to have direct anthelmintic effect on gastrointestinal nematodes (see Chap. 2). They can interact with rumen bacteria in ruminants but their direct immunomodulatory effect on mucosal and/or systemic immunity in infected animals has been little studied.

3.3 Natural Compounds with Antioxidant and Immunomodulatory Effects: Synergism with Anthelmintic Drugs

We showed in the previous chapters that the variety of chemicals present in plant extracts or invertebrate organisms have different pharmacological activities such as anthelmintic, antimicrobial, antioxidant, and anti-inflammatory, which may act in a synergistic manner resulting in the overall clinical effect (Gurib-Fakim 2006). These activities have been implicated for phenolic compounds including condensed tannins and flavonoids as well as saponins, terpenoid portions of essential oils and alkaloids (Makkar et al. 2007). Most plant-derived secondary metabolites are capable of directly affecting inflammatory mediators, as well as the production/activity of second messengers, transcription factors, and key proinflammatory molecule expressions (Calixto et al. 2003, 2004). Phytochemicals with phenolic structures have the dominant position in the list of metabolites and, of these, flavonoids attracted probably the widest interest assessed by a high number of reviews and book chapters (for example: Andersen and Markham 2006). The structural differences between the different flavonoid classes relate to the chemistry of C ring in the molecule, variation in the number and distribution of the phenolic hydroxyl groups across the molecules, and their substitutions. Despite close structural characteristics, the biological and biochemical properties vary considerably with only minor modifications in structure. The number of phenolic hydroxyl groups, the extent and nature of the substitutions, and their specific positions on the ring structure influence whether they function as modulators of enzyme activity or as antioxidant, cytotoxic, or antimutagenic agents in vitro and in vivo (Rice-Evans 2004).

Reactive oxygen species (ROS) are implicated in numerous pathophysiological events such as aging, cancer, atherosclerosis, and diabetes. Natural antioxidants are reported to provide substantial protection that slows down the process of oxidative damage caused by ROS. The central role of ROS in mediating the pathology in many diseases has stimulated interest in the possible role of natural antioxidant agents in preventing the development of these diseases. There is an emerging view that flavonoids, and their in vivo metabolites, do not act as conventional hydrogen-donating antioxidants, but may exert modulatory actions on cells through actions at protein kinase and lipid kinase signaling pathways (Williams et al. 2004). In parasitic infections, overproduction of ROS by activated neutrophils, macrophages, and eosinophils is a nonspecific effector mechanism called the respiratory burst, which is aimed to kill protozoan or damage the metazoan parasites inside inflammatory lesion. On the cellular level, free radicals generated during oxidative stress can have deleterious effects on macromolecules causing peroxidation of cell lipids, DNA fragmentation, and protein oxidation. The chronic inflammation associated with the continuous oxidative stress in the tissues was shown to participate in initiation of fibrogenesis in the liver (Baroni et al. 1998; Reeves and Friedman 2002).

Many developmental stages of helminths, especially trematodes and cestodes, persist in the liver of the mammalian host, where they contribute to the development of fibrosis. A severe pathology in schistosomiasis is due to formation of liver egg granulomas and secondary hepatic fibrosis. The larval (metacestode) stage of cystic and alveolar alveococcosis is characterized by a granulomatous host reaction and fibrogenesis surrounding parasitic cysts, which are believed to protect the host against the parasite growth (Guerret et al. 1998).

Current drugs used to combat these infections (praziquantel, albendazole, or tribendimidine, respectively) are not able to attenuate the oxidative stress in the tissue directly, but rather they can modulate or decrease the activity of hosts antioxidant systems. We showed that praziquantel decreased levels of the important nonenzymatic antioxidant molecule glutathione and did not reduce ROS-stimulated fibrogenesis in the livers with the metacestode stage of *Mesocestoides vogae* infection for several days (Velebný et al. 2010). The effect of albendazole on antioxidant enzymatic systems, the glutathione transferases (GST), was examined in an experimental nematode infection caused by *Trichinella spiralis* in muscle (Wojtkowiak et al. 2007a) and in the liver (Wojtkowiak et al. 2007b). It is known that GSTs are the major enzymes participating in glutathione metabolism. Authors of these studies showed that albendazole treatment mediated changes in the quaternary structure of cytosolic GST and in vitro inhibited about six times the total activity of these enzymes at temperatures above 30 °C.

In experimental schistosomiasis treated with praziquantel, therapy caused not only significant reduction in the worm numbers, but interestingly, also the partial resorption of collagens and amelioration of granuloma pathology (Singh et al. 2004). Effects seen at the molecular level were decreased expression of profibrotic factors Tumor necrosis factor α (TNF-α), Transforming growth factor β (TGF-β) and iNOS gene expression in the livers of mice, which developed within several months' post therapy. The reduction of fibrosis cannot occur when pathogenic stimuli in the tissues persist, which is the case of *Schistosoma* reinfections or liver infections with asexually developing metacestode stages of several cestodes species. Complementary treatment is an alternative approach, when the conventional drug is combined with other compounds having a different mechanism of action. Such an alternative was shown to be effective in the treatment of schistosomiasis, when short- or long-term administration of the antifibrotic drugs, natural products, or melatonin have been suggested to be potential complements to the ordinary therapy (El-Shenawy et al. 2008; El-Ridi et al. 2010; Rabia et al. 2010).

In view of large body of information about medicinal applications of secondary plant metabolites, it is evident that some compounds exert antiparasitic as well as immunomodulatory activities, while others can act synergistically with drugs only as antioxidant and immunomodulatory agents. In this respect, only a few chemically defined secondary metabolites have been examined on experimental parasitic infections to date. *Paeoniflorin* is a phenolic compound and is one of the major constituents of herbal medicine derived from plant *Paeonia lactiflora* Pall. (Fig. 3.2a, b). It has been reported to have immunoregulatory and anti-inflammatory effects but its antiparasitic activity has not been evaluated yet, probably

Fig. 3.2 a,b. Image of plant *Paeonia lactiflora* Pall., the roots of which are the source of phenolic compound paeoniflorin and its structure

Paeoniflorin

due to no records about parasiticidal effect of this plant in ethnomedicine. Chu et al. (2007) demonstrated that this compound has potent antifibrotic effect in vitro. As was mentioned above, chronic ROS production and fibrogenesis are interconnected processes. Oxidative stress in hepatocytes contributes to hepatic stellate cell (HSC) activation indirectly by releasing profibrogenic factors, among which transforming growth factor $\beta 1$ (TGF-$\beta 1$) is the main HSC activator (De Blesser et al. 1999). Kupffer cell (liver macrophages) infiltration and activation also contributes to HSC activation through the actions of cytokines (especially TGF-$\beta 1$) and released ROS (Bilzer et al. 2006), resulting in the production of collagens. In the study of Chu et al. (2007) soluble antigens released by *Schistoma japonicum* eggs were potent activators of macrophages, which responded by elevated secretion of TGF-$\beta 1$. Paeoniflorin in a dose-dependent manner inhibited TGF-$\beta 1$ mediated collagen production from HSC in vitro. Recently, Chu et al. (2011) confirmed the antifibrotic effect of this compound in vivo in mice with *S. japonicum* infection, showing that paeoniflorin administration indirectly and directly suppressed the alternative activation of macrophages through decreasing secretion of IL-13. Interleukin (IL)-13 and alternatively activated macrophages play an important role in liver granuloma formation and fibrosis in schistosomiasis, and therefore this phenolic natural compound might be used as prophylactic agent for these pathologies.

Flavonoids comprise a great variety of compounds and the isoflavone silymarin has attracted the attention of several groups of parasitologists. The flavonoid silymarin, isolated from the seeds of milk thistle (*Silybum marianum* L. Gaertn.), has a high human acceptance, since it is used around the world for the treatment of liver diseases and its hepatoprotective properties have been reported by many authors. It can selectively induce apoptosis in cancer cells while it has no effect on normal cells in the liver or other tissues (Zhong et al. 2006; Zheng et al. 2011; Ramasamy and Agarwal, 2008). It has been discovered that silymarin can act as (i) antioxidant, scavenger of free radicals in damaged tissue, and regulator of intracellular content of GSH; (ii) cell membrane stabilizer and permeability regulator; (iii) promoter of ribosomal RNA synthesis and proteosynthesis (for example: Dvořak et al. 2003; Fraschini et al. 2002). Silymarin consists of four flavonolignan isomers namely: silybin (also known as silibinin), isosilybin, silydianin, and silychristin (Kroll et al. 2007) (Fig. 3.3a, b). Silibinin is the most prevalent

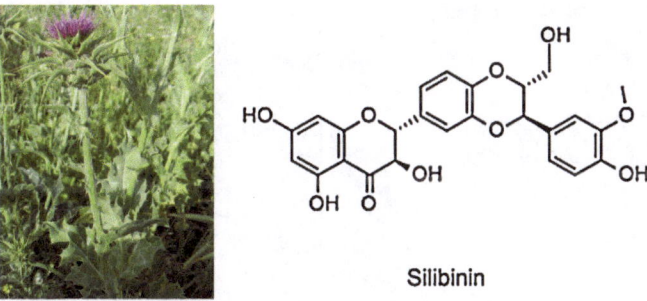

Silibinin

Fig. 3.3 a,b. Image of plant *Silybum marianum,* the seeds of which are the source of silymarin isoflavonoides and structure of the silibinin

and biologically active of the four isomers and represents about 60–70 % of silymarin, followed by silychristin (20 %), silidianin (10 %), and isosilybin (5 %).

In the study by Mata-Santos et al. (2010) on the effect of silymarin treatment in experimental schistosomiasis, silymarin did not affect parasite oviposition capacity; but reduced granulomatous peri-ovular reaction in the liver, and decreased hepatic fibrosis in this infection. The mechanism of antifibrotic effect of silymarin administration was further studied on the same model infection by El-Lakkany et al. (2012). Treatment of *S. mansoni*-infected mice with silymarin alone at 4 and 12 weeks post infection resulted in remarkable worm and egg reduction accompanied by a partial increase in the percentage of dead eggs 10 and 18 weeks post infection. This was associated with healing of hepatic granulomatous lesions as evidenced histopathologically. Authors suggested that silymarin eliminated the products of oxidative reactions and assisted in the immune-mediated destruction of worms and eggs. In addition, treatment with silymarin, whether in the acute or chronic stage of infection, significantly reduced the hepatic collagen content, tissue expression of TGF-β1, and matrix metalloproteases 2 (MMP-2) and the number of mast cells. During follow-up the therapy, it restored levels of glutathione, which was depleted due to the liver damage by infection. Administration of silymarin in addition to PZQ resulted in the complete eradication of schistosome worms, presence of nonviable eggs, and reduction in the hepatic tissue egg loads. Administration of silymarin with PZQ did not interfere with, or affect, the antischistosomal activity of PZQ. On the contrary, combined treatment showed enhanced efficacy and caused the highest reduction in TGF-β1 and MMP-2 levels and the lowest mast cell numbers in granulomatous lesions (El-Lakkany et al. 2012).

The direct trematocidal effect in vitro has not been studied to date but was not suggested by authors. We carried out a series of in vitro tests on larval stage of cestode *Mesocestoides vogae,* where we showed that incubation of larvae with silymarin dissolved in DMSO up to concentration of 0.03 mg/ml medium had no larvicidal effect (Fig. 3.4). Larvae retained the typical motility and all of them showed macroscopic changes in contrast with larvae incubated with 0.03 µg/ml praziquantel, which killed all larvae. A direct in vivo parasiticidal effect

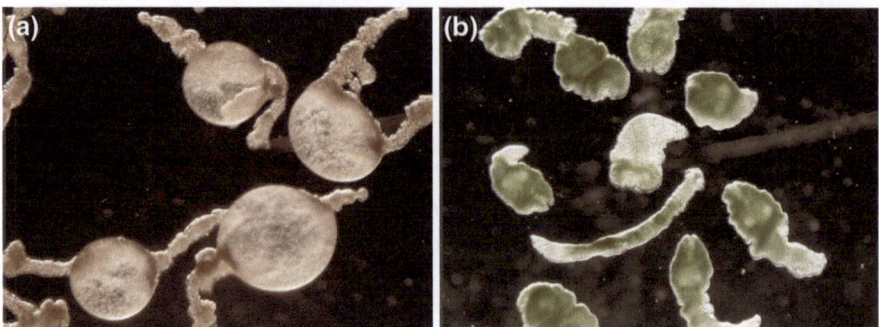

Fig. 3.4 In vitro effect of praziquantel (0.3 μg/ml) and silymarin (0.3 mg/ml) on Mesocestoides vogae larvae after 48 h of incubation

of silymarin was not demonstrated in dogs with asymptomatic *Giardia* infection (Chon and Kim 2005). Giardiosis is a protozoan infection with worldwide distribution that can cause clinically important diseases in gastrointestinal tracts of humans and animals. The number of parasitic cysts in fecal samples of silymarin-treated dogs was the same as in the control group, indicating no parasiticidal effects. However, the combined therapy with drug metronidazole plus silymarin was more effective (91 %) in comparison with that of metronidazole given alone (75 %), showing significant synergistic effect of silymarin probably as antioxidant and immunomodulatory compound.

The larval stage (tetrathyridium) of the cestode *Mesocestoides vogae* (syn. *M. corti*) multiplies asexually in the liver and peritoneal cavity of mice and was recommended as a suitable experimental model for slowly developing human metacestode infections in pharmacological and biological studies (WHO 1995). Migration and multiplication of larvae cause severe damage to the liver parenchyma, which results in oxidative stress, hepatocyte dysfunctions, severe inflammation, and progressing fibrogenesis. Another site of larval multiplication is the peritoneal cavity, where ascites and a massive accumulation of inflammatory cells take place. Due to the similar pathology and immune responses with medically important cestodes, we employ this model to get some insights into antifibrotic and immunomodulatory effects of silymarin as the possible adjunct to albendazole or praziquantel therapy. Single therapy with this flavonoid mixture was ineffective against infection and rather resulted in the higher larval numbers in the livers as the result of reduced encapsulation of larvae in collagenous layers. However, in our further study (Hrčkova and Velebný, 2010) we showed that co-administration of praziquantel and silymarin (SIL) to infected mice significantly enhanced the efficacy of the treatment in the livers in comparison with administration of the drug alone (Table 3.1). Intensity of fibrogenesis was quantified by means of determination of hydroxyproline (HP) concentrations in the livers. Whereas PZQ-therapy resulted in higher HP concentrations in the comparison with the livers from untreated mice, combined therapy with PZQ plus SIL resulted in a significantly

Table 3.1 Efficacy (%) of antihelmintic drug praziquantel (PZQ) administered alone or in combination with flavonoid silymarin (SIL) on *Mesocestoides vogae* larvae, non-encapsulated or surrounded by the fibrous capsules (encapsulated) in the livers of mice (Hrčkova and Velebný 2010)

Days	Praziquantel		Praziquantel+silymarin	
p.i./p.t.	Non-encapsulated (%)	Encapsulated (%)	Non-encapsulated (%)	Encapsulated (%)
25/1	34.4	25.0	47.8[a]	39.0[a]
28/4	23.1	47.7	36.9[a]	42.8
35/11	10.9	43.6	31.4[a]	69.5
44/20	0	58.2	2.7	78.4

Legend [a] Significantly higher efficacy after PZQ + SIL versus PZQ treatment ($P < 0.05$) for non-encapsulated larvae significantly higher efficacy after PZQ + SIL versus PZQ treatment ($P < 0.05$) for encapsulated larvae

suppressed fibrosis level. On the molecular level, collagen type I expression was elevated following administration of PZQ, but co-administration of SIL resulted in a temporary downregulation of gene expression. Collagen type III gene expression was only slightly downregulated following PZQ therapy but significantly decreased after combined therapy with PZQ and SIL. A strong antioxidant effect of silymarin was confirmed in vitro (Hrčkova and Velebný 2010) as well as in vivo during the follow-up combined therapy (Velebný et al. 2010) in line with observations of other authors in different experimental models of fibrosis. Using the same infection and treatment design, the enzymes ALT and AST (serum markers of liver damage) and hyaluronic acid (serum marker of ongoing fibrogenesis) were significantly leveled down after the combined therapy (Velebný et al. 2008). Moreover, we showed that treatment with PZQ and silymarin prevented lipid peroxidation, stimulated GSH synthesis, and proliferation of hepatocytes in infected livers. Elimination of reactive oxygen species in the course of combined therapy resulted also in the significant decrease of total collagen synthesis in the liver (Fig. 3.5a, b) as well as mastocytosis, and modulated the number/proportion of granulocytes in the inflammatory lesions. In the light of the above findings we proposed that potentiation of the larvicidal activity of praziquantel with silymarin is mediated via its antioxidant effect, resulting in the downregulation of fibrogenesis and consequently in higher availability of drug for the parasite. Ascites associated with peritonitis and eosinophilia in mice with *M. vogae* infection seemed to be a good model to evaluate immunomodulatory activity of various compounds like polysacharide glucan (Hrčkova and Velebný 2007) and silymarin (Hrčkova and Velebný S 2012). In the livers, eosinophils and neutrophils predominated in the acute inflammatory lesions, where they probably contributed to the activation of hepatic stellate cells by means of high amount of released reactive oxygen species (ROS). With the progressing infection, most of the polymorphonuclear leucocytes were replaced by fibroblasts and mast cells, the numbers of which gradually elevated in the granulomatous fibrous lesions, suggesting their role in the perpetuation of fibrogenesis. In the livers, immunomodulatory activity of silymarin and praziquantel co-administration resulted in the significant decline

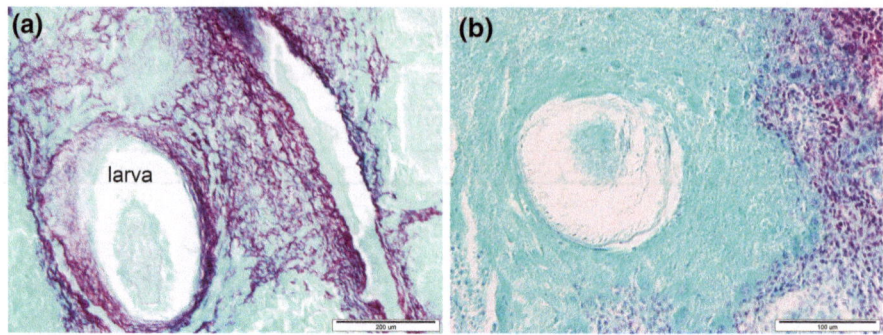

Fig. 3.5 a,b. Histochemical staining of total collagen on liver sections (Picrosirius *Red*/Fast *Green*) in the liver from mice with *Mesocestoides vogae* infection, which were treated with praziquantel alone (**a**) or with the combination of praziquantel and silymarin (**b**). Collagen layer surrounding larva and migratory track were thicker in the livers of PZQ-treated mice than in livers of mice after combined therapy

in mast cell numbers, downregulation of IL-5, IL-4 expression and stimulation of IFN-γ gene expression, indicating the restoration of Th1/Th2 balance in favor of antifibrotic Th1 type of responses. The significant potentiation of the praziquantel larvicidal effect in the peritoneal cavities following co-administration of silymarin was also demonstrated (Table 3.2) along with the elevated gene expression of IFN-γ and TNF-α pro-inflammatory cytokines in exudate cells. The proportions of inflammatory cell types were also significantly modulated toward a marked decline of alternatively activated macrophages by PZQ and PZQ plus silymarin-therapy. In addition, treatments resulted in a higher proportion of neutrophils and/or lymphocytes, respectively, suggesting that antibody-dependent larvicidal effect of praziquantel (Fallon et al. 1996) was further enhanced by coadministration of silymarin.

Similarly as was found for other flavonoids, the direction of silymarin immunomodulatory activity is influenced by the disease-specific pathology. At least two following studies provided some evidence for such a hypothesis. In one study silibinin, a component of SIL mixture polarized the Th1/Th2 immune balance

Table 3.2 Efficacy (%) of antihelmintic drug praziquantel (PZQ) administered alone or in combination with flavonoid silymarin (SIL) on *Mesocestoides vogae* larvae in the peritoneal cavities of mice

Days	Efficacy of therapy in peritoneal cavity of mice	
p.i./p.t.	Praziquantel	Praziquantel+silymarin
25/1	38.1 ± 3.8	70.2 ± 4.7*
28/4	39.4 ± 5.4	73.6 ± 8.3*
35/11	57.6 ± 6.6	80.5 ± 5.8*
44/20	51.0 ± 5.2	66.1 ± 6.6

Legend * Significantly higher efficacy after PZQ + SIL versus PZQ treatment ($P < 0.05$)

toward the Th1 direction in association with the increased production of IFN- γ in an experimental model of IgE-mediated allergy (Kuo and Jan 2009). Another study reported on the inhibition of CD4 + proliferation, decrease of IFN-γ, and IL-2 production after incubation of splenocytes from healthy mice with silymarin (Gharagozloo et al. 2010). Table 3.3 summarizes list of plants and their botanical families used in the studies cited in Chap. 2 and 3.

3.4   Concluding Remarks

Most plant-derived secondary metabolites are capable of directly affecting inflammatory mediators, as well as the production/activity of second messengers, transcription factors, and key pro-inflammatory molecule expressions in mammals. Phytochemicals with phenolic structures have the dominant position in the list of metabolites and, of these, flavonoids attracted probably the widest interest as many studies suggested multidirectional action and very low toxicity. Several phenolic compounds have been shown to interfere also with helminth-specific physiological pathways, namely artemisinins, genistein, curcumin, and monomers of tannins. Therefore, combined therapy would take an advantage of the synergistic action of a drug and a natural compound which exert antiparasitic as well as immuno-modulatory activities.

Co-administration of strong antioxidant phenolic molecules is an effective approach when parasite-stimulated liver fibrogenesis is also present, which can be downregulated as well as mastocytosis. In this respect, two plant-derived compounds, paeoniflorin and silymarin, did not show the direct vermicidal effects on cestodes and trematodes. Their co-administration with praziquantel significantly potentiated efficacy of treatment and contributed to the normalization of the host pathophysiology and immunosuppresion, offering a promising alternative treatment option for chronic parasitic diseases.

3.5   Appendix

Table 3.3 Alphabetical list of plants and their families with anthelmintic effects cited in this book

Name of plant tested for anthelmintic activity	Family	Type of material investigated:extract/ isolated compounds
Acacia oxyphylla	Mimosaceae	+ / +
Acacia auriculiformis	Mimosaceae	+ / +
Adhatoda vasica	Acathaceae	+ / -
Aegle marmelos Corr	Rutaceae	+ / -
Agave sisalama	Asparagaceae	+ / +

(continued)

Table 3.3 (continued)

Name of plant tested for anthelmintic activity	Family	Type of material investigated:extract/ isolated compounds
Ageratum conyzoies	Asteraceae	+ / +
Ailanthus altissima	Simaroubaceae	- / +
Allium sativum	Amaryllidaceae	+ / +
Anacardium humile	Anacardiaceae	+ / +
Anethum graveolens	Apiaceae	- / +
Ananas comosus	Bromeliaceae	+ / +
Annona squamosa	Annonaceae	+ / -
Arachis pintoi	Fabaceae	+ / -
Artemisia absinthiun	Asteraceae	+ / +
Artemisia annua	Asteraceae	+ / +
Artemisia vulgaris	Asteraceae	+ / +
Artemisia herba-alba	Asteraceae	+ / +
Baccharis dracunculifolia	Asteraceae	- / +
Beta vulgaris	Amaranthaceae	- / +
Bidens sulphurea	Asteraceae	- / +
Caesalpinia crista	Fabaceae	+ / -
Caesalpinia bonducella	Fabaceae	+ / -
Calendula officinalis	Asteracae	- / +
Carica papaya	Caricaceae	- / +
Chenopodium ambrosioides	Amaranthaceae	- / +
Chicorium intybus	Asteraceae	+ / +
Cinnamomum camphora	Lauraceae	+ / +
Cocos nucifera	Arecaceae	+ / +
Coriandrum sativum	Apiaceae	+ / +
Corydalis turtschaninovani	Papaveraceae	- / +
Croton zehntneri	Euphorbiaceae	- / +
Curcuma longa	Zigiberaceae	+ / +
Cymbopogon schoenanthus	Poaceae	- / +
Cymbopogon martini	Poaceae	- / +
Dracaena fragrans	Agavaceae	- / +
Dryopteris filix-max	Dryopteridaceae	+ / -
Dorycnium spp	Fabaceae	+ / +
Eucalyptus staigeriana	Myrtaceae	+ / +
Eucalyptus citriodola	Myrtaceae	+ / -
Ficus glabrata	Moraceae	- / +
Flemingia vestita	Fabaceae	+ / +
Fumaria parviflora	Papaveraceae	+ / -
Hedera helix	Araliaceae	+ / -
Hedysarum coronariun	Fabaceae	+ / +
Chelidomium majus	Papaveraceae	- / +
Lantana camara	Verbenaceae	+ / -
Lysimachia ramosa	Primulaceae	+ / -
Lespedeza cuneata	Fabaceae	+ / -
Leucosidea sericea	Rosaceae	+ / -

(continued)

Table 3.3   (continued)

Name of plant tested for anthelmintic activity	Family	Type of material investigated:extract/ isolated compounds
Lippia sidoides	Verbenaceae	+ / -
Lotus corniculatus	Nelumbonaceae	+ / -
Lotus pedunculatus	Nelumbonaceae	+ / -
Macleaya cordata	Papaveraceae	- / +
Medicago sativa	Fabaceae	+ / +
Melia azedarach	Meliaceae	+ / -
Mentha piperita	Lamiaceae	+ / +
Newbouldia laevis	Bignoniaceae	+ / -
Nigella sativa	Ranunculaceae	- / +
Ocimum gratissimum	Lamiaceae	+ / -
Olea europaea	Oleaceae	+ / +
Onobrychis vicifolia	Fabaceae	+ / +
Origanum compactum	Lamiaceae	- / +
Paeonia lactiflora Pall.	Ranunculaceae	+ / +
Picrasma quassioides	Simaroubaceae	- / +
Piper tuberculatum	Piperaceae	+ / +
Piper cubeba	Piperaceae	+ / +
Piper longum	Piperaceae	- / +
Piper tuberculatum	Piperaceae	- / +
Phytolacca icosandra	Phytolaccaceae	+ / -
Plectranthus neochilus	Lamiaceae	- / +
Prosopsis laevigata	Fabaceae	+ / -
Salvia spp.	Lamiaceae	- / +
Satureja khuzestanic	Lamiaceae	+ / -
Salvadora persica	Salvadoraceae	+ / -
Sericea lespedeza	Fabaceae	+ / -
Stephania glabra	Menispermaceae	- / +
Silybum marianum	Asteraceae	- / +
Solanum myriacanthum	Solanaceae	+ / -
Solanum torvum	Solanaceae	+ / -
Spigelia anthelmia	Loganiaceae	+ /-
Streblus asper	Moraceae	- / +
Strobilanthes discolor	Acanthaceae	+ / -
Struthiola argentea	Thymelaeaceae	- / +
Styrax camporum	Styraceae	- / +
Styrax pohlii	Styraceae	- / +
Taminalia avicennoides	Combrataceae	+ / -
Trachyspermum ammi	Apiaceae	- / +
Trifolium repens	Fabaceae	+ / -
Trichilia claussenii	Meliaceae	+ / -
Vitex negundo	Lamiaceae	+ / -
Zanthoxylum rhetsa DC	Rutaceae	+ / -
Xylocarpus granatum	Meliaceae	- / +

References

Aggarwal BB, Harikumar KB (2009) Potential therapeutic effects of curcumin, the anti-inflammatory agent, against neurodegenerative, cardiovascular, pulmonary, metabolic, autoimmune and neoplastic diseases. Int J Biochem Cell Biol 41:40–59. doi:10.1016/j.biocel.2008.06.010

Agrawal DK, Mishra PK (2010) Curcumin anti cancer action. Med Res Rev 30:818. doi:10.1002/med.20188

Allam G (2009) Immunomodulatory effects of curcumin treatment on murine schistosomiasis mansoni. Immunobiol 214:712–727. doi:10.1016/j.imbio.2008.11.017

Andersen QM, Markham KR (2006) Flavonoids: chemistry, biochemistry and applications. Taylor and Francis Group, Boca Raton

Anthony RM, Rutizky LI, Urban JF, Stadecker MJ, Gause WC (2007) Protective immune mechanisms in helminth infection. Nat Rev Immunol 7:975–987. doi:10.1038/nri2199

Araujo CAC, Leon LL (2001) Biological activities of *Curcuma longa L.* Mem Inst Oswaldo Cruz 96:723–728. doi:org/10.1590/S0074-02762001000500026

Balint GA (2001) Artemisinin and its derivatives: an important new class of antimalarial agents. Pharmacol Ther 90:261–265. doi:org/10.1016/S0163-7258(01)00140-1

Baroni GS, D'Ambrosio L, Farretti G, Casini A, Di Sario A, Salzano R, Ridolfi F, Saccomanno S, Jezequel AM, Benedetti A (1998) Fibrogenic effect of oxidative stress on rat hepatic stellate cells. Hepatology 27:720–726. doi:10.1002/hep.510270313

Bilzer M, Roggel F, Gerbes AL (2006) Role of Kupffer cells in the host defense and liver disease. Liver Int 26:1175–1186. doi:10.1111/j.1478-3231.2006.01342.x

Burke ML, Jones MK, Gobert GN, Li YS, Ellis MK, McManus DP (2009) Immunopathogenesis of human schistosomiasis. Parasit Immunol 31:163–176. doi:10.1111/j.1365-3024.2009.01098.x

Calixto JB, Otuki MF, Santos AR (2003) Anti-inflammatory compounds of plant origin. Part I. Action on arachidonic acid pathway, nitric oxide and nuclear factor kappa B (NF-kappa B). Planta Med 69:973–983. doi:10.1055/s-2003-45141

Calixto JB, Campos MM, Otuki MF, Santos AR (2004) Anti-inflammatory compounds of plant origin. Part II. Modulation of pro-inflammatory cytokines, chemokines and adhesion molecules. Planta Med 70:93–103. doi:10.1055/s-2004-815483

Chon SK, Kim NS (2005) Evaluation of silymarin in the treatment on asymptomatic *Gairdia* infections in dogs. Parasitol Res 97:445–451. doi:10.1007/s00436-005-1462-z

Chu D, Luo Q, Li C, Gao Y, Yu L, Wei W, Wu Q, Shen J (2007) Paeoniflorin inhibits TGF-β1-mediated collagen production by *Schistosoma japonicum* soluble egg antigen in vitro. Parasitology 134:1611–1621. doi:10.1017/S0031182007002946

Chu D, Du M, Hu X, Wu Q, Shen J (2011) Paeoniflorin attenuates schistosomiasis japonica-associated liver fibrosis through inhibiting alternative activation of macrophages. Parasitology 138:1259–1271. doi:10.1017/S0031182011001065

Das R, Roy A, Ganguly A, Datta N, Majumder HK (2008) Curcumin, a dietary polyphenol, emerges as a novel inhibitor of DNA topoisomerase I of kinetoplastid parasite *Leishmania donovani.* Biochem J. doi:10.1042/BJ20081134

De Blesser PJ, Xu G, Romboust K, Rogiers V, Geerts A (1999) Glutathione levels discriminate between oxidative stress and transforming growth factor-β signalling in activated rat hepatic stellate cells. J Biol Chem 274:33881–33887. doi:10.1074/jbc.274.48.33881

Dixon JB (1997) Echinococcosis. Comp Immun Microbiol Infect Dis 20:87–94. PII: S0147-9571(96)00019-7

Duchen MR (2000) Mitochondria and calcium: from cell signalling to cell death. J Physiol (Lond) 529:57–68. doi:10.1111/j.1469-7793.2000.00057.x

Dvořák Z, Kosina P, Walterova D, Šimanek V, Bachleda P, Ulrichova J (2003) Primary cultures of human hepatocytes as a tool in toxicity studies: cell protection against model toxins by flavonolignans obtained from *Silybum marianum*. Toxicol Lett 137: 201–212. doi:PII: S0378-4274(02)00406-X

Eckstein-Ludwig U, Webb RJ, van Goethem IDA, East JM, Lee AG, Kimura M, O'Neill PM, Bray PG, Ward SA, Krishna S (2003) Artemisinins target the SERCA of *Plasmodium falciparum*. Nature 424:957–961. doi:10.1038/nature01813

Efferth T, Dunstan H, Sauerbrey A, Miyachi H, Chitambar CR (2001) The anti-malarial artesunate is also active against cancer. Int J Oncol 18:767–773

El-Lakkany NM, Hammam OA, El-Maadawy WH, Badawy AA, Ain-Shoka AA, Ebeid FA (2012) Anti-inflammatory/anti-fibrotic effects of the hepatoprotective silymarin and the schistosomicide praziquantel against *Schistosoma mansoni*-induced liver fibrosis. Parasites & Vectors 5: art.no.9. doi:10.1186/1756-3305-5-9

El-Ridi R, Aboueldahab M, Tallima H, Salah M, Mahana N, Fawzi S, Mohamed SH, Fahmy OM (2010) In vitro and in vivo activities of arachidonic acid against *Schistosoma mansoni* and *Schistosoma haematobium*. Antimicrob Agents Chemother 54:3383–3389. doi:10.1128/AAC.00173-10

El-Shenawy NS, Soliman MF, Reyad SI (2008) The effect of antioxidant properties of aqueous garlic extract and *Nigella sativa* as anti-schistosomiasis agents in mice. Rev Inst Med Trop Sao Paulo 50:29–36. doi:org/10.1590/S0036-46652008000100007

Fallon PG, Fookes RE, Wharton GA (1996) Temporal differences in praziquantel- and oxamniquine-induced tegumental damage to adult *Schistosoma mansoni*: implication for drug-antibody synergy. Parasitology 112:47–58. doi:org/10.1017/S0031182000065069

Fischer M, Regitz Ch, Kahl M, Werthebach M, Boll M, Wenzel U (2012) Phytoestrogens genistein and daidzein affect immunity in the nematode *Caenorhabditis elegans* via alterations of vitellogenin expression. Molec Nutr & Food Res 56:957–965. doi:10.1002/mnfr.201200006

Fraschini G, Demartini G, Esposti D (2002) Pharmacology of silymarin. Clin Drug Invest 22:51–65

Freitas TC, Pearce EJ (2010) Growth factors and chemotactic factors from parasitic helminths: molecular evidence for roles in host-parasite interactions versus parasite development. Int J Parasitol 40:761–773. doi:10.1016/j.ijpara.2010.02.013

Fujiwara N, Kobayashi K (2005) Macrophages in inflammation. Curr Drug Targets Inflamm Allergy 4:281–286

Gao F, Wei D, Bian T, Xie P. Zou J, Mu H, Zhang B, Zhou X (2012) Genistein attenuated allergic airway inflammation by modulating the transcription factors T-bet, GATA-3 and STAT-6 in a murine model of asthma. Pharmacology 89: 229–236. doi:10.1159/000337180

Gause WC, Urban Jr JF, Stadecker MJ (2003) The immune response to parasitic helminths: insights from murine models. Trends Immunol 24:269–277. doi:10.1016/S1471-4906(03)00101-7

Gharagozloo M, Velardi E, Bruscoli S, Agostini M, Di Sante M, Donato V, Amirghofran Z, Riccardi C (2010) Silymarin suppress CD4$^+$T cell activation and proliferation: Effects on NF-$_\kappa$B activity and IL-2 production. Pharmacol Res 61:405–409

Golenser J, Waknine JH, Krugliak M, Hunt NH, Grau GE (2006) Current perspectives on the mechanism of action of artemisinins. (Review article). Int J Parasitol 36:1427–1441. doi:10.1016/j.ijpara.2006.07.011

Gottstein B, Hemphill A (2008) *Echinococcus multilocularis*: the parasite-host interplay. Exp Parasitol 119:447–452. doi:10.1016/j.exppara.2008.03.002

Guerret S, Vuitton DA, Liance M, Pater C, Carbillet JP (1998) *Echinococcus multilocularis*: relationship between susceptibility/resistance and liver fibrogenesis in experimental mice. Parasitol Res 84:657–667. doi: not found

Gurib-Fakim A (2006) Medicinal plants: traditions of yesterday and drugs of tomorrow. Mol Aspects Med 27:1–93. doi:10.1016/j.mam.2005.07.008

Haddad M, Sauvain M, Deharo E (2011) Curcuma as a parasiticidal agent: a review. Planta Med 77(672–678):21104602. doi:10.1055/s-0030-1250549

Hämäläinen M, Nieminen R, Vuorela P, Heinonen M, Moilanen E (2007) Anti-inflammatory effects of flavonoids: genistein, kaempferol, quercetin, and daidzein inhibit STAT-1 and

NF-$_\kappa$B activations, whereas flavone, isorhamnetin, naringenin, and pelargonidin inhibit only NF-$_\kappa$B activation along with their inhibitory effect on iNOS expression and NO production in activated macrophages. Mediators Inflam Article ID: 45673, 10 pp doi:10.1155/2007/45673

Hemaiswarya S, Kruthiventi AK, Boble M (2008) Synergism between natural products and antibiotics against infectious diseases. Phytomedicine 15:639–652. doi:10.1016/j.phymed.2008.06.008

Hewitson JP, Grainger JR, Maizels RM (2009) Helminth immunoregulation: the role of parasite secreted proteins in modulating host immunity. Mol Biochem Parasitol 167:1–11. doi:10.1016/j.molbiopara.2009.04.008

Houghton PJ, Howes MJ, Lee CC, Steventon G (2007) Uses and abuses of in vitro tests in ethnopharmacology: visualizing an elephant. J Ethnopharmacol 110:391–400. doi:10.1016/j.jep.2007.01.032

Hrčkova G, Velebný S (2007) Antibody response in mice infected with *Mesocestoides vogae* (syn. *Mesocestoides corti*) tetrathyridia after treatment with praziquantel and liposomised glucan. Parasitol Res 100:1351–1359. doi:10.1007/s00436-006-0434-2

Hrčkova G, Velebný S (2010) Flavonoid silymarin potentiates anthelmintic effect of praziquantel via down-regulation of oxidative stress and fibrogenesis in the liver. In: Proceedings of the World Medical Conference, Malta, Sept 15–17 2010. WSEAS Press, Wisconsin, pp 250–257

Hrčkova G, Velebný S G (2012) Current situation and new possibilities in pharmacology of parasitic infections. Proceedings of the World Medical Conference, Kos-Island, July 14–17 2012, Wisconsin, WSEAS Press, pp 106–112

Kang JS, Park KH, Lee H, Park KH, Kim HM (2010) Artemisinin inhibits lipopolysaccharide-induced nitric oxide production by blocking IFN-β production and STAT-1 signaling in macrophages. J Immunol 184:142.2 doi: not found

Keiser J, Utzinger J (2007) Artemisinins and synthetic trioxolanes in the treatment of helminth infections. Curr Opin Infect Dis 20:605–612. doi:10.1097/QCO.0b013e3282f19ec4

Keiser J, N'Guessan NA, Adoubryn KD, Silué KD, Vounatsou P, Hatz CH, Utzinger J, N'Goran K (2010) Efficacy and safety of mefloquine, artesunate, mefloquine-artesunate and praziquantel against *Schistosoma haematobium*: randomized, exploratory open-label trial. Clin Inf Dis 50:1205–1215. doi:10.1086/651682

Kolodziej H, Kayser O, Kiderlen AF, Hideyuki ITO, Hatano T, Yoshida T, Foo LY (2001) Proanthocyanidins and related compounds: Antileishmanial activity and modulatory effects on nitric oxide and Tumor Necrosis factor-α-release in the murine macrophage-like cell line RAW 264.7. Biol Pharm Bull 24:1016–1021. doi:10.1248/bpb.24.1016

Krishna S, Bustamante L, Haynes RK, Staines HM (2008) Artemisinins: their growing importance in medicine (Review). Trends Pharmacol Sci 29:520–527. doi:10.1016/j.tips.2008.07.004

Kroll DJ, Shaw HS, Oberlies NH (2007) Milk thistle nomenclature: why it matters in cancer research and pharmacokinetic studies. Integr Cancer Ther 6:110–119. doi:10.1177/1534735407301825

Kuhn I, Kellenberger E, Said-Hassane F, Villa P, Rognan D, Lobstein A, Haiech J, Hibert M, Schuber F, Muller-Steffner H (2010) Identification by high-throughput screening of inhibitors of *Schistosoma mansoni* NAD$^+$ catabolizing enzyme. Bioorg Medic Chem 18:7900–7910. doi:10.1016/j.bmc.2010.09.041

Kuo FH, Jan TR (2009) Silibinin attenuates antigen-specific IgE production through the modulation of Th1/Th2 balance in ovalbumin-sensitized BALB/c mice. Phytomedicine 16:271–276. doi:10.1016/j.phymed.2008.07.006

Lightowlers MW, Rickard MD (1988) Excretory–secretory products of helminth parasites: effects on host immune responses. Parasitology 96:S123–S166. doi:org/10.1017/S0031182000086017

Liu R, Dong HF, Guo Y, Zhao QP, Jiang MS (2011) Efficacy of praziquantel and artemisinin derivatives for the treatment and prevention of human schistosomiasis: a systematic review and meta-analysis. Parasites & Vectors 4: Article ID 201. doi:10.1186/1756-3305-4-201

Magalhães LG, Machado CB, Morais ER, Moreira EB, Soares CS, da Silva SH, da Silva Filho AA, Rodrigues V (2009) In vitro schistosomicidal activity of curcumin against *Schistosoma mansoni* adult worms. Parasitol Res 104:1197–1201. doi:10.1007/s00436-008-1311-y

Maheshwari RK, Singh AK, Gaddipati J, Srimal RC (2006) Multiple biological activities of curcumin: a short review. Life Sci 78:2081–2087. doi:10.1016/j.lfs.2005.12.007

Maizels RM, Yazdanbakhsh M (2003) Immune regulation by helminth parasites: cellular and molecular mechanisms. Nat Rev Immunol 3:733–744. doi:10.1038/nri1183

Makkar HPS, Francis G, Becker K (2007) Bioactivity of phytochernicals in some lesser-known plants and their effects and potential applications in livestock and aquaculture production systems. Animal 1:137–139. doi:10.1017/S1751731107000298

Mata-Santos HA, Lino FG, Rocha CC, Paiva CN, Castelo Branco MT, Pyrrho Ados S (2010) Silymarin treatment reduces granuloma and hepatic fibrosis in experimental schistosomiasis. Parasitol Res 107:1429–1434. doi:10.1007/s00436-010-2014-8

Meschnik SR (2002) Artemisinin: mechanisms of action, resistance and toxicity. Int J Parasitol 32:1655–1660. PII:S0020-7519(02)00194-7

Mishra S, Karmodiya K, Surolia N, Surolia A (2008) Synthesis and exploration of novel curcumin analogues as anti-malarial agents. Bioorg Med Chem 16:2894–2902. doi:10.1016/j.bmc.2007.12.054

Mota MLR, Thomas G, Filho BJM (1985) Anti inflammatory actions of tannins isolated from the bark of *Anacardium occidentale* L. J Ethnoparmacol 13:289–300. doi:10.1016/0378-8741(85)90074-1

Nery PS, Nogueira FA, Martins ER, Duarte ER (2010) Effects of *Anacardium humile* leaf extracts on the development of gastrointestinal nematode larvae of sheep. Vet Parasitol 171:361–364. doi:10.1016/j.vetpar.2010.03.043

Olliaro PL, Haynes RK, Meunier B, Yuthavong Y (2001) Possible modes of action of the artemisinin-type compounds. Trends Parasitol 17:122–126. PII:S1471-922(00)01838-X

Perkins S, Verschoyle RD, Hill K, Parveen I, Threadgill MD, Sharma RA, Williams ML, Steward WP, Gescher AJ (2002) Chemopreventive efficacy and pharmacokinetics of curcumin in the min/+ mouse, a model of familiar adenomatous polyposis. Cancer Epidemiol Biomarkers Preven 11:535–540

Polkowski K, Mazurek AP (2000) Biological properties of genistein. Review of in vitro and in vivo data. Acta Polon Pharmac—Drug Res 57:135–155

Rabia I, Nagy F, Aly E, Mohamed A, EL-Assal F, El-Amir A (2010) Effect of treatment with antifibrotic drugs in combination with PZQ in immunized *Schistosoma mansoni* infected murine model. J Am Sci 6:208–216

Ramasamy K, Agarwal R (2008) Multitargeted therapy of cancer by silymarin. Cancer Lett 269:352–362. doi:10.1016/j.canlet.2008.03.053

Reeves HL, Friedman SL (2002) Activation of hepatic stellate cells—a key issue in liver fibrosis. Front BioSci 7:D808–D826

Rice-Evans C (2004) Flavonoids and isoflavones : absorption, metabolism, and bioactivity (Serial Review) (Ed. Rice-Evans). In: Flavonoids and isoflavones (phytoestrogens): absorption, metabolism, and bioactivity. Free Radical Biol Med 36:827–828. doi:10.1016/j.freeradbiomed.2003.12.012

Rothenberg ME, Hogan SP (2006) The eosinophils. Anu Rev Immunol 24:147–174

Sattelle DB, Buckingham SD, Akamatsu M, Matsuda K, Pienaar I, Jones AK, Sattelle BM, Almond A, Blundell CD (2009) Comparative pharmacology and computational modelling yield insights into allosteric modulation of human α7 nicotinic acetylcholine receptors Biochem Pharmacol 78:836–843. doi:10.1016/j.bcp.2009.06.020

Sen R, Ganguly S, Saha P, Chatterjee M (2010) Efficacy of artemisinin in experimental visceral leishmaniasis. Int J Antimicrob Agents 36:43–49. doi:10.1016/j.ijantimicag.2010.03.008

Shakir L, Hussain M, Javeed A, Ashraf M, Riaz A (2011) Artemisinins and immune system. (Review). Eur J Pharmacol 668:6–14. doi:10.1016/j.ejphar.2011.06.044

Sing KP, Gerard HC, Hudson AP, Boros DL (2004) Expression of matrix metalloproteinases and their inhibitors during the resorption of schistosome egg-induced fibrosis in praziquantel-treated mice. Immunology 111:343–352. doi:10.1111/j.I365-2567.2004.01817.x

Sutherland IA, Lee DL (1993) Acetylcholinesterase in infective-stage larvae of *Haemonchus contortus*, *Ostertagia circumcincta* and *Trichostrongylus colubriformis* resistant and susceptible to benzimidazole anthelmintics. Parasitology 107:553–557. doi:10.1017/S003118200006813X

Tandon V, Pal P, Roy B, Rao HSP, Reddy KS (1997) In vitro anthelmintic activity of root-tuber extract of *Flemingia vestita*, an indigenous plant in Shillong, India. Parasitol Res 83:492–498. doi:10.1007/s004360050286

Taylor CK, Levy RM, Elliott JC, Burnett BP (2009) The effect of genistein aglycone on cancer and cancer risk: a review of in vitro, preclinical and clinical studies. Nutr Rev 67:398–415. doi:10.1111/j.1753-4887.2009.00213.x

van Riet E, Hartgers FC, Yazdanbakhsh M (2007) Chronic helminth infections induce immunomodulation: consequences and mechanisms. Immunobiology 212:475–490. doi:10.1016/j.imbio.2007.03.009

Vanoevelen J, Dode L, Van Baelen K, Fairclough RJ, Missiaen L, Raeymaekers L, Wuytack F (2005) The secretory pathway Ca^{2+}/Mn^{2+}-ATPase 2 is a Golgi localized pump with high affinity for Ca^{2+} ions. J Biol Chem 280:22800–22808. doi:10.1074/jbc.M501026200

Velebný S, Hrčkova G, Kogan G (2008) Impact of treatment with praziquantel, silymarin and/or b-glucan on pathophysiological markers of liver damage and fibrosis in mice infected with *Mesocestoides vogae* (Cestoda) tetrathyridia. J Helminthol 82:211–219. doi:10.1017/S0022149X08960776

Velebný S, Hrčkova G, Königová A (2010) Reduction of oxidative stress and liver injury following silymarin and praziquantel treatment in mice with *Mesocestoides vogae* (Cestoda) infection. Parasitol Int 59:524–531. doi:10.1016/j.parint.2010.06.012

Verdrengh M, Collins LV, Bergin P, Tarkowski A (2004) Phytoestrogen genistein as an anti-staphylococcal agent. Microbes Infect 6:86–92. doi:10.1016/j.micinf.2003.10.005

Wang J, Zhou H, Zheng J, Cheng J, Liu W, Ding G, Wang L, Luo P, Lu Y, Cao H, Yu S, Li B, Zhang L (2006) The antimalarial artemisinin synergizes with antibiotics to protect against lethal live *Escherichia coli* challenge by decreasing proinflammatory cytokine release. Antimicrob Agents Chemother 50:2420–2427. doi:10.1128/AAC.01066-05

Wang J, Zhang Q, Jin S, He D, Zhao S, Liu S (2008) Genistein modulate immune responses in collagen-induced rheumatoid arthritis model. Maturitas 59:405–412. doi:10.1016/j.maturitas.2008.04.003

WHO (1995) Guidelines for surveillance, prevention, and control of echinococcosis/hydatidosis, 2nd edn. World Health Organisation, Geneva

Williams RJ, Spencer JPE, Rice-Evans K (2004) Flavonoids: antioxidants or signalling molecules? (Serial Review) (Ed. Rice-Evans) In: Flavonoids and isoflavones (phytoestrogens): absorption, metabolism, and bioactivity. Free Radical Biol Med 36: 838–849. doi:10.1016/j.freeradbiomed.2004.01.001

Wojtkowiak A, Boczoń K, Wandurska-Nowak E (2007a) Effect in vitro of albendazole on the kinetics of cytosolic glutathione transferase from the rat liver. Wiad Parazytol 53:97–102

Wojtkowiak A, Boczoń K, Wandurska-Nowak E, Derda M (2007b) Evaluation of effects of albendazole on the kinetics of cytosolic glutathione transferase in skeletal muscles during experimental trichinellosis in mice. Parasitol Res 100:647–651. doi:10.1007/s00436-006-0285-x

World Health Organization (2005) Strategic orientation paper on prevention and control of malaria, roll back malaria department. Available at. www.who.int/malaria/docs/trainingcourses/NPOreport.pdf. Accessed May 11 2006

Zhao M, Xue DB, Zheng B, Zhang WH, Pan SH, Sun B (2007) Induction of apoptosis by artemisinin revealing the severity of inflammation in caerulein-induced acute pancreatitis. World J Gastroenterol 13:5612–5617

Zheng D, Wang Y, Zhang D, Liu Z, Duan C, Jia L, Wang F, Liu Y, Liu G, Hao L, Zhang Q
(2011) In vitr o antitumor activity of silybin nanosuspension in PC-3 cells. Cancer Lett
307:158–164. doi:10.1016/j.canlet.2011.03.028

Zhong X, Zhu Y, Lu Q, Zhang J, Ge Z, Zheng S (2006) Silymarin causes caspase activation and
apoptosis in K562 leukemia cells through inactivation of Akt pathway. Toxicology
29:211–216. doi:10.1016/j.tox.2006.07.021